Rethinking Green Politics

Sage Politics Texts

Series Editor
IAN HOLLIDAY
University of Manchester

SAGE Politics Texts offer authoritative and accessible analyses of core issues in contemporary political science and international relations. Each text combines a comprehensive overview of key debates and concepts with fresh and original insights. By extending across all main areas of the discipline, SAGE Politics Texts constitute a comprehensive body of contemporary analysis. They are ideal for use on advanced courses and in research.

Rethinking Green Politics

Nature, Virtue and Progress

John Barry

SAGE Publications
London • Thousand Oaks • New Delhi

First published 1999

SAGE Publications Ltd
6 Bonhill Street
London EC2A 4PU

SAGE Publications Inc.
2455 Teller Road
Thousand Oaks, California 91320

SAGE Publications India Pvt Ltd
32, M-Block Market
Greater Kailash – I
New Delhi 110 048

British Library Cataloguing in Publication data

A catalogue record for this book is available
from the British Library

ISBN 0 7619 5605 0
ISBN 0 7619 5606 9 (pbk)

Library of Congress catalog card number 98–61591

Typeset by Mayhew Typesetting, Rhayader, Powys
Printed in Great Britain by Biddles Ltd, Guildford, Surrey

Contents

Acknowledgements

I dedicate this book to my partner Yvonne who has not only been an unfailing source of encouragement, solace and support, but has also commented extensively upon it and made numerous suggestions. Her patience and forbearance (and the occasional demand to stop prevaricating!) have been instrumental and invaluable in writing this book. For constantly reminding me that there is more to life than simply reflecting upon it, I am particularly grateful.

Like many first books, this one began life as a PhD thesis. As such, I would like to thank my supervisor, Chris Berry, to whom I owe a great debt of gratitude for his patience and 'wise stewardship'. In particular I wish to thank him for his excellent supervision sessions, for extensive and detailed comments on every draft, and most of all for reminding me that while academic work may be qualitatively different from other forms of employment, it is after all a job. I thank him for reminding me of the dangers of taking oneself too seriously. The coherence and clarity of the book owe much to him. It goes without saying that any shortcomings I claim for myself.

Thanks are also due to my colleagues at Keele who have commented on or with whom I have discussed parts of my argument. My colleagues in the Politics Department – Margaret Canovan, Paula Casal, Andy Dobson, Brian Doherty, John Horton, Rosemary O'Kane and Lucy Sargisson (now at Nottingham) – may find some of the points they raised either included or answered in the text. I would also like to thank colleagues in other departments at Keele, particularly Mat Paterson in International Relations and John Proops in Environmental Social Sciences. Thanks are also due to those undergraduate students who took my Nature and Politics course from 1994 to 1996, those who took the MA in Environmental Politics (1997–8), and finally the lively and convivial research community on green political issues within both the Department of Politics and the University more widely.

The 1992 European Consortium on Political Research Workshop on Democracy and Green Politics in Madrid provided me not only with an opportunity to present my ideas to some of those thinkers, writers and commentators whose work peppers this book, but also with the opportunity to debate and argue with them. Sections of Chapter 7 were written for and developed from that workshop, and I would like to take this opportunity to thank all the participants. Marius de Geus, who also attended that workshop, had already very kindly commented on some early drafts of Chapters 4, 5 and 7, and I am grateful for his generosity. The Green Politics Reading Group at Keele, where I presented an early version of Chapter 2, also deserves a word of thanks. Other individuals that I would like to thank include Marcel Wissenburg for his extensive and witty comments on an earlier draft of the book, Kay Milton whose anthropological insights were particularly useful in Chapter 3, Avner de-Shalit and Graham Smith, Damian White and all those who attended workshops at Interdisciplinary Research Network on Environment and Society (IRNES) conferences from 1992 to 1995. Ian Holliday and anonymous referees offered suggestions which have improved the book, and, together with Lucy Robinson at Sage, were instrumental in enabling me to 'de-PhD' the manuscript. Responsibility for any enduring marks of the latter on the book is entirely mine.

Finally, I would like to thank all those I have not mentioned with whom I have discussed, argued and debated over the last number of years, both within and without academia. Long may the arguments continue.

John Barry
Keele

1

Introduction

CONTENTS

Although it would be an exaggeration to proclaim that we are all greens now, it is true to say that green politics has become an established perspective in political and moral debates, both within the academy and within society at large. Green politics in this sense is no longer 'green', in the sense of 'immature', and nor is a concern with the environment the exclusive province of self-proclaimed greens. In recent years, the 'mad cow' disease crisis, public fears over genetic engineering, concerns over global warming, and a host of other social–environmental issues demonstrate the increasing sensitivity of democratic populations to environmental risks. Such dilemmas, their occurrence, frequency and public perception, also consolidate green politics, its values, principles and concerns, as an established analytical perspective on contemporary democratic politics. A critical analysis of the political theory underpinning this perspective is the aim of this book.

This book sets out to examine current understandings of green political theory, and presents an alternative conceptualization, one which, I argue, accords better with its central informing principles and values. The aim of the book is to rethink green politics by focusing more on green political *theory* than on green political *ideology*. That is, it tries to develop an alternative understanding of green politics which is, at times, markedly at odds with dominant views of green politics which are profoundly ideological in character. For example, the conception of green political theory outlined and defended here takes issue with the ecocentric and anarchistic tendencies which are taken as central in defining green ideology (Dobson, 1990; 1995; Eckersley, 1992a; Goodin,

1992). Instead a naturalistic but anthropocentric moral base (Chapter 3) is argued to underpin an ecological political arrangement termed 'collective ecological management' (Chapter 5), a central component of which involves the *transformation* as opposed to the *abolition* of the liberal democratic state. One major difference between green political theory and green political ideology is therefore the modesty of the former.[1] While both green ideology and theory share a common critical stance, they differ in that green political theory is less marked by a utopian style of critique. In rethinking green politics I argue that an immanent, as opposed to a utopian, critique should characterize green political theory. This idea of immanent critique is a key methodological approach in developing the book's argument, and immanence is held to be important for rethinking green politics in two respects. Immanent critique here simply means a preference for working from within the existing conceptual or real-world situation towards an alternative understanding or position, as opposed to basing critique on a 'view from nowhere'. On the one hand, the conceptualization of green theory which is offered is itself the product of an immanent, i.e. internal, critique and reconstruction of existing conceptualizations of green ideology and theory. On the other hand, the particular conceptualization of green political theory defended is one for which the resolution of social–environmental problems must at least start from an immanent (as opposed to an 'external') critique of existing patterns and modes of social–environmental interaction.

The critical aspects of this book are not ends in themselves, but ought to be read as a necessary part of the process of developing a more coherent, plausible and attractive version of green political theory. These critical aspects are part of the general attempt to reconstruct green political theory via a process of immanent critique. The alternative understanding of green politics developed in the book comes from within green political theory, broadly understood. Thus, many 'sacred cows' of green political theory are questioned, such as ecocentrism (Chapters 2 and 3), a principled anti-state position (Chapters 4 and 5), anti-urbanism (Chapter 4), post-materialism (Chapter 6), an antipathy to market economic relations (Chapter 6), and a bias towards direct democracy (Chapter 7). The concern is not to dispose of these sacred cows (an act that would be singularly problematic given the vegetarian and vegan predilections of many greens!), but rather to interrogate them in order to extract a defensible kernel, and then to integrate the latter into an alternative understanding of green political theory. This critical-reconstructive aim is part and parcel of developing a green political theory from within the various discourses of green political ideology. The aim is to critically assess the main understandings and accounts of green ideology and translate the principles and values they embody into an alternative political idiom. For example, deep ecology's ecocentrism is understood as an argument concerning the necessity to regulate and reformulate anthropocentric moral reasoning. Ecocentrism, on this gloss, is a warning against a complacent and potentially arrogant anthropocentrism, and suggests the reform of anthropocentric moral reasoning through the introduction of the idea of 'ecological virtue'. As will be argued in Chapters 2 and 3, the normative claims of green political theory do not require the rejection of anthropocentric moral reasoning in favour of a putative non-anthropocentric ecocentrism. Likewise eco-anarchism (discussed in Chapter 4) is viewed as a 'regulative' as opposed to a 'constitutive' ideal of green politics. That is, eco-anarchism expresses the green concern for democratization, decentralization and appropriate scale, and acts as a reminder of the ecological and democratic dangers of centralized and hierarchical political authority and economic organization.

While this interpretation of dominant understandings of green moral and political theory will doubtless be criticized, even rejected, by those who conceive of green politics in terms of 'deep/radical' versus 'shallow/reformist' (that is, a profoundly *ideological* perspective), the alternative understanding of green political theory which emerges from this book cannot be rejected on the grounds that the political theory it defends is not 'green' in a generic sense. The argument of the book is thus premised on a rejection of understandings of green politics based around distinguishing 'deep' from 'shallow' green thinking, or 'ecologism' from 'environmentalism'. These accounts of green politics, while of course important and valuable, often obscure as much as they reveal. This ideological approach tends to highlight the differences rather than the connections, the overlapping principles and themes between various conceptualizations of normative green theory. As will be suggested in Chapter 4, such ideological views of green politics were perhaps an inevitable aspect of its early development, but are now detrimental to its future development. Looking back on the development of green politics since the 1960s, one is keenly aware of how its critical, radical and often uncompromising character underpins this ideological account.

Green Ideology versus Green Political Theory

Recent work on green political thought has opened out the theoretical space to allow for less ideological accounts to be developed. One indication of this is recent scholarship which tries to articulate various combinations of existing political ideologies with ecologism, such as green politics and liberalism (Wissenburg, 1998), eco-socialism (Pepper, 1993), eco-Marxism (O'Connor, 1995; Harvey, 1996; Barry, 1998d) and eco-feminism (Plumwood, 1993). The aim of distinguishing green ideology – ecologism – from green political theory is not to devalue ideological accounts as mere polemic, or to claim that such writings lack any genuine theoretical merit. Rather the distinction is used to carve out another area of green political space, one which in large extent would not be possible without previous debates within the ideological framework of recent green theorizing. Perhaps one mark of whether this book is successful in its aims is whether it will be criticized by existing schools of green politics – social ecology, deep ecology, eco-socialism etc. – either as a misrepresentation of green politics or as abstract academic theorizing and therefore not worth considering.

Broadly speaking ideological accounts of green politics differ from theoretical ones in terms of their respective styles of theorizing. Ideological accounts of green politics, for example, tend to be characterized by listing the 'core principles of green politics' in opposition to some other ideological position (Porritt, 1984), or 'non-true' green positions (O'Riordan, 1981; Paehlke, 1989; Pearce, 1992). Theorizing green politics is not as restricted as mapping the ideological territory of green politics; that is, green political theory ranges over a larger theoretical area than green political ideology. In a sense green political theory is not committed to a particular 'party line' and can thus engage in theoretical explorations without having to worry about adhering to certain fixed or *a priori* principles or values. Above all else it recognizes the dynamic nature of political theorizing in respect to social–environmental relations, and does not accept that representatives of the 'green movement' have a monopoly on thinking about those relations.

Dobson's book, one of the most influential academic books on green politics, while entitled *Green Political Thought* (1990; 1995), could more accurately be described as an analysis of green political ideology. His distinction between 'environmentalism' and 'ecologism' was largely motivated by a concern to explore the 'real' or 'true' green ideological position. 'Ecologism' was the 'true' green political viewpoint, while 'environmentalism' was not. Ecologism is held to be marked by its radicalism which is made up of, *inter alia*, an acceptance of the 'limits to growth' hypothesis, an eco-anarchist critique of present political and economic arrangements, a particular vision of the future 'sustainable society', a moral critique of anthropocentrism and an endorsement of ecocentrism. Thus, he and others who also employed this binary and ideological understanding of green politics sought to emphasize the novelty and difference of green concerns from other political ideologies such as liberalism, socialism and conservatism. This concern to emphasize the differences between 'ecologism' and existing political ideological positions went hand in hand with the need to stress differences within green politics between 'true' and 'false' accounts. However, as I have argued elsewhere (Barry, 1994), such binary, ideological accounts of green politics are unnecessarily restrictive, and can become a hindrance to the future evolution of green politics. Such ideological accounts tend to become dogmatic and inflexible and to produce 'closure' rather than 'openness' and 'space' in theoretical debate and development. My aim of rethinking green politics takes as a starting point Hayward's conclusion in his recent study that

> If ecological politics has to be radical to be ecological (versus environmentalism), it also has to be realistic (versus ecologism). The challenge for a green political theory is to avoid the fallacies of ecologism without lapsing back into the complacencies which are attendant on a reformist environmentalism. (1995: 200)

In terms of the modesty of theoretical as opposed to ideological accounts of green politics mentioned above, my argument is that if green political theory is to have a 'utopian' complexion, then this ought to be understood as referring to what Hayward (1995) calls a 'concrete utopia' or what Harvey (1996) calls a 'utopianism of process', to be distinguished from the 'closed', blueprint-like and abstract utopian visions of ecologism. According to Lukes, concrete utopia depends on the 'knowledge of a self-transforming present, not an ideal future' (1984: 158). Thus green politics as the search for concrete utopias fits with the immanent-reconstructive approach adopted in this book.

Green political theory is ecologically based, but not ecologically centred. This is meant in two senses. Firstly, green political theory is not solely concerned with environmental issues and social–environmental relations, but speaks across a full range of normative concerns. Secondly, green political theory takes existing ideological conceptualizations of green thought as its starting point, but is not limited to the range of issues articulated by these conceptualizations. Thus while taking on board some of the topics raised within ideological accounts, green political theory is marked by an attempt to both critically assess the putative 'green' position on these topics, and if necessary challenge and move beyond what are seen as 'basic positions' of green thought. The basic premise for a theoretical account of green politics is to admit that 'ecological' thought does not have a monopoly on defining the scope, principles or values of green politics.

Ideological accounts of green politics are characterized by a tendency to neglect the difficult task of working out the theoretical and practical implications of their principles and values. Green ideology, in common with most other ideologies, assumes the harmony of its principles by positing a future social order in which these principles are realized (Bookchin, 1980; Carter, 1993). Ideological accounts of 'ecologism' focus on describing the 'sustainable society' to the neglect, for example, of working out the implications of the principle of sustainability (Dobson, 1990; 1995). Working out, and through, the principles and values of green politics is a primary aim here. As such, the book seeks to establish the rationality and persuasiveness of the green case independently of the attractiveness of green visions of the 'sustainable society'. One of the problems with ideological interpretations of green politics is the 'external' and often abstract quality of their diagnostic and prescriptive elements. They offer a 'view from nowhere' as a guide to get from 'here' (the unsustainable present) to 'there' (the future 'sustainable society').[2] Thus, for example, deep ecology's (ideological) view of the 'ecological crisis' as a *crisis of* Western culture is contrasted with a view of ecological problems as a *contradiction within* that culture (as will be shown in Chapter 2). As a crisis of Western culture, the deep ecological 'solution' is premised on an external critique: the rejection of anthropocentrism and the adoption of ecocentrism. If anthropocentrism, which is a core cultural orientation of Western societies, is the cause of the 'ecological crisis', then an immanent critique of anthropocentrism is clearly insufficient. Nothing short of a cultural 'paradigm shift' and a 'new ecocentric ethic' is required, since Western culture and anthropocentrism cannot resolve the crisis they have caused. This understanding is a major reason why deep ecology can be viewed as a 'redemptive politics', for which an ecocentric 'reverence for nature' as opposed to a 'respect for nature' is a necessary condition for the resolution of the ecological crisis (Chapter 2). This ideological view is rejected on the grounds that the ecological crisis is not a crisis of civilization in the way some radical greens think, but is better viewed as a cultural contradiction which is resolvable from within the resources of Western, anthropocentric culture. That is, to adequately understand the ensemble of social–environmental problems, dilemmas and risks faced by contemporary Western societies, one must analyse this ensemble in terms of its *causes* rather than just its *effects*. In short, one must look for the causes of social–environmental problems *within* society and culture first, in order to properly address their effects as manifested in problematic relations *between* the human and the non-human worlds.

Recent developments within green political thought, to which this book is intended to contribute, indicate a certain 'maturing' of green politics, or what Hayward has termed 'theoretical consolidation' (1995: 6). This has been marked by a shift from criticizing the status quo and advocating a moral 'paradigm shift' and/or a vision of the future 'sustainable society', to a concern with formulating feasible and attractive solutions, policies and institutional designs to present social–environmental dilemmas. For example, much of the recent work on green political theory is concerned with spelling out the normative and practical implications of ecological sustainability (Norton, 1991; Barry and Proops, 1996; Doherty and de Geus, 1996; Jacobs, 1996; Dobson, 1998).

One of the aims of this book is to demonstrate that translating green normative and political principles into policies does not necessarily rob green politics of its radical and transformative character. While utopian visions of a future better society and ecocentric forms of moral reasoning are transgressive and contribute enormously to the imaginative and innovative spirit of green thinking, it is a mistake to think that only these forms of thinking are 'radical'. Concrete utopian theorizing can be equally radical. It is also just as radical (if not more) to transcend the dualistic viewpoint of conventional understandings of green politics. One of the reasons for adopting a 'critical' or 'immanent-reconstructive' approach stems from a belief that the theoretical consolidation and development of green politics are now as much about getting rid of the unnecessary as about developing additional insights. The critical-reconstructive approach can be viewed as an attempt to put the green theoretical house in order, as it were. The elaboration of an alternative conceptualization of green theory is a necessary prelude in order that it take its proper place in contemporary debates within political theory. Part of this immanent critique involves elaborating key terms and principles used in political theory from a green perspective. Thus alongside discussing 'green' principles such as 'sustainability' and green concerns with establishing what I call 'morally symbiotic' relations with the non-human world, green political theory is also characterized by its particular understanding of standard political theory issues and debates around 'liberty' (Chapter 6), 'interests' and 'preferences' (Chapters 3, 5, 6, 7), 'the state' (Chapters 4, 5, 7), 'progress' (Chapters 7, 8) and 'democracy' (Chapter 7).

One suggested area of exploration in this search for a theoretical account of green politics is to focus on the types of policies that greens, of various hues, do or should endorse, given their values and principles. This is in keeping with the aim suggested above which focuses on areas of overlap. Often, in practice, the difference between 'radicals', 'ecocentrics' and so-called 'shallow ecologists' or 'reformists' is one of degree rather than kind. This is particularly the case with policy proposals concerning environmental protection and preservation. What one often finds is substantive agreement on policies or institutional reform but disagreement on the reasons given for supporting policies. Thus, one of the major arguments of the book is that there is a large area of practical agreement between different conceptualizations of green ideology. For example, in respect to the 'ecocentric–anthropocentric' dichotomy, following Norton's (1991) 'convergence hypothesis', I argue that a reformed 'naturalistic humanism' (Chapter 3) can support a 'stewardship ethic' (Chapters 3, 5, 7), which can further integrate green demands for symbiotic and sustainable relations between human societies and their environments (Chapter 3). This stewardship position is normatively based on the notion of 'ecological virtue' and an 'ethics of use' for the environment in which human interests and non-human interests can be harmonized (but never completely). The advantage of the stewardship position is that it is politically (as well as philosophically) superior to ecocentrism, since it holds that care for the environment cannot be independent of human interests.

Ecological stewardship, unlike ecocentrism, seeks to emphasize that a self-reflexive, long-term anthropocentrism, as opposed to an 'arrogant' or 'strong' anthropocentrism, can secure many of the policy objectives of ecocentrism, in terms of environmental preservation and conservation. As argued in Chapter 3, a reformed, reflexive anthropocentrism is premised on critically evaluating human uses of the non-human world, and distinguishing 'permissible' from 'impermissible' uses. That is, an 'ethics of use', though anthropocentric and rooted in human interests, seeks to regulate human interaction with the environment by distinguishing legitimate 'use' from unjustified 'abuse'. The premise for this defence of anthropocentric moral reasoning is that an immanent critique of 'arrogant humanism' is a much more defensible and effective way to express green moral concerns than rejecting anthropocentrism and developing a 'new ecocentric ethic'. As discussed in Chapters 2 and 3, ecocentric demands are premised on an over-hasty dismissal of anthropocentrism which precludes a recognition of the positive resources within anthropocentrism for developing an appropriate and practicable moral idiom to cover social–environmental interaction.

A central part of developing an alternative language for green political theory involves a concern with 'virtue' and 'progress'. Although no specific chapters are devoted to either of these, they are constant points of reference throughout. A concern with 'ecological virtue' is a recurrent theme throughout the book, from Chapter 2 where it is suggested as an alternative ethical idiom by which to express ecocentric moral concerns, to Chapter 7 where it is used to integrate green democratic concerns of citizenship with ecological stewardship. Picking up on the deep ecological point about the centrality of character to morality in Chapter 2, a virtue ethics approach is used throughout the book as a way of establishing the connection between green moral and political theory. As developed in Chapter 3, a virtue ethics approach views the moral dimension of green concerns as having less to do with finding the correct set of moral rules by which we are to interact with nature, than with cultivating respectful, less 'arrogant' modes of interaction with, and understandings of, the non-human world. Another advantage of a virtue ethics approach to green moral and political theory is that central to virtue is the idea of flourishing, or well-being, rather than simply 'welfare'. This concern to shift the moral focus of green concerns to the area of 'flourishing' as opposed to the economistic overtones of 'welfare' (usually understood to mean income or material consumption) fits, as argued later in Chapter 6, with the green critique of economic growth.

Virtue ethics can also furnish a much needed sense of proportion and humility to guard against the vices of hubris, indifference and disrespect towards the non-human world. Through the cultivation of ecologically sensitive modes of relating to the world, particularly human transformative relations and practices, the change that greens argue for can thus acquire a cultural as well as a political character. This cultural dimension is stressed throughout the book as another way in which the green political theory I defend is to be distinguished from other forms. Thus throughout reference is made to human modes of action and interaction such as 'production', 'consumption' and 'citizenship' which are central to social–environmental relations. Green politics, centred on the cultivation of 'ecological stewardship', becomes a matter of integrating these modes of interaction so that together they constitute a stewardship mode. Taking a virtue-orientated position implies that a concern of green politics is to create modes of human interaction with the non-human world which are ecologically sustainable and morally symbiotic. The latter refers to fostering self-reflexive modes of human behaviour in which human interests, for which particular human uses of the non-human world are carried out, are considered as necessary but not sufficient to justify that use. Green political theory thus becomes concerned with discriminating legitimate, worthy or serious human interests from illegitimate, unworthy or trivial ones. The concern with virtue is thus related to discriminating 'symbiotic' from 'parasitic' human modes of interaction, and to foster the former as a (ecologically) virtuous mode. At the same time, the classical view of virtue as a mean between extremes is also evident in the reflexive form of anthropocentrism developed in Chapter 3. This reformed anthropocentrism is a mean between the extremes (vices) of deep ecological 'submissiveness' in respect to nature, and the 'arrogance' of 'strong anthropocentrism' or the 'domination of nature'.[3]

In a similar fashion the discussion of 'progress' also shadows the book. While specific issues around progress, understood as 'economic development' and 'modernization', are directly addressed in Chapters 6 and 7 respectively, the force of the term 'progress' in the title is intended as a guide to how the book should be read. This focus on progress is in keeping with the reconstructive intention of the argument, suggesting that green politics is not anti-progress, anti-modern or anti-Enlightenment. Rather green political theory is to be understood as an immanent critique of progress, suggesting an alternative understanding of it based on its view of social and social–environmental relations. One reason why progress is chosen as a pivotal issue, through which green political theory is discussed, is a conviction that what green politics represents is analogous to the challenges and opportunities that marked the Enlightenment or the advent of 'modernity'. In common with recent writers on green politics, I argue that the central aims of green politics coalesce around the necessity and desirability of 'ecological enlightenment' (Beck, 1995). Thus the discussion of 'progress' is intended not simply to reassert the 'progressive' credentials of green politics (Paehlke, 1989), but to suggest that the explicit examination of what constitutes 'progress' is central to the task of rethinking green politics.

In rethinking green politics a primary aim of this book is to invite debate and discussion on future developments within green theory and practice. Much of the tone, and indeed contents, of the book are due to a certain sceptical attitude I have regarding many green claims, ideas and proposals, which should not be read, however, as being dismissive of them. Rather, in the spirit of debate and discussion, I feel strongly that such a sceptical frame of mind is absolutely essential in ascertaining the worth, veracity, desirability or necessity of any political theory, particularly when, as in the case of green political theory, its implementation may have tremendous, and as yet only dimly perceived, effects on individual, social, political and economic life.

This scepticism is however tempered with a humility in that I do not believe a fully rethought green politics is to be found within these pages. In such a multifaceted and rapidly changing area as green politics, which at times seems to leave no aspect of the 'human condition' untouched, it would be arrogant and foolish to presume to have surveyed the whole terrain and mapped a new and 'true' course for its future development. The logic of the argument in this book is that 'rethinking green politics' is (or ought to be seen as) a continual process, indeed a constitutive component of the development of green politics. While the comfortable certainties of ideology have their attractions, political theory cannot be satisfied with such potential intellectual indolence. One reason for this is that the stakes are too high. Even if we reject, as I suggest, the apocalyptic view of an 'ecological crisis' facing humanity, there are enough real social–environmental problems and dilemmas to make the duty of accurately conceptualizing them for their political implications more than simply an abstract, academic task. Unlike Marx's famous injunction to the workers of the world, the environmental problems facing us are such that there is more than chains and human oppression at stake.

Notes

---

1 This modesty of theoretical as opposed to ideological conceptions of green politics refers not only to the fact that non-utopian critiques and political alternatives tend to envisage less radical change, but also to the Western-centred character of the account developed here. While the reconceptualization of green politics advanced here may have some purchase in non-Western contexts, any claim for the indiscriminate or universal validity of the proposed conception is emphatically *not* intended.

2 See Mellor (1995) for a critical analysis and reversal of the common green

strategy of getting from 'here' (the unsustainable present) to 'there' (the future sustainable society). According to her, there are central aspects of 'here' (notably within feminized spheres of social life) which are sustainable, while 'there' represents the unworkable utopianism of market-led productive abundance. Thus the point is not to get from 'here' to 'there', but to stay 'here' and resist 'there'. Also see Salleh (1997).

3 While the ecological vice of deep ecology can be seen to rest on its failure to adequately differentiate humanity from nature, the ecological vice of arrogant anthropocentrism is based on a false separation of humanity from the rest of the natural order. From one end, the similarities between humanity and nature are over-emphasized, while from the other they are denied. As a mean between two extremes (vices), an ecological virtue approach adopts a mode of apprehension and interaction with the natural world in which humans are seen as *a part of* but also *apart from* nature (Barry, 1995c).

2

Rethinking Green Ethics I: From Deep Ecology to Ecological Virtue

CONTENTS

This chapter looks at deep ecology as the pre-eminent ecocentric approach within green moral theory. The aim of this chapter is not to offer a comprehensive overview of deep ecology, but to argue that as it stands deep ecology is insufficient to ground green political claims and policy prescriptions. The basic argument is that deep ecology is unable to provide the necessary normative basis for green political theory. A central reason for this failure is that deep ecology's non-anthropocentrism is premised on a false understanding of anthropocentrism. Allied to this is the particular understanding of morality and ethics within deep ecology, an understanding which gives little attention to the collective, intersubjective character of the ethical as a sphere of human action. The aim of this chapter is to clear the ground for the argument in the next where it is argued that anthropocentric moral reasoning is not only perfectly legitimate but fundamentally necessary to green politics if the gap between its political and philosophical claims is to be overcome.

The starting point for this critique of deep ecology is given by Dobson who points out that there is a rupture between ecophilosophy (by which he means deep ecology) and green politics, concluding that 'the politics of ecology does not follow the same ground rules as its philosophy' (1990: 68). That is, to all intents and purposes they seem to be two independent discourses which are contingently rather than inherently connected. This lack of normative coherence between the political and the moral level is of course a serious deficiency. So long as deep ecology is considered as the moral basis of green politics, this unhelpful and unnecessary division will persist. This chapter will seek to provide reasons why we must seriously question the view that 'There must be no doubt that Deep Ecology is indeed the Green Movement's philosophical basis' (Dobson, 1989: 41) or that 'ecocentrism' is the 'correct' normative underpinning for green politics (Eckersley, 1992a: 26–31). What I wish to demonstrate is that ultimately the ecocentric/anthropocentric division is a false and damaging dichotomy which severs the continuity between green moral and political theory. Part of this discontinuity has to do with the difficulty of securing political agreement on the basis of the substantive metaphysical commitments that characterize deep ecology. In other words, if deep ecology is the normative core of green political theory then it may actually undermine the political relevance of green politics, in terms of securing democratic agreement for green policies.

An overall aim then of this chapter is to highlight the fact that the normative basis of green political theory consists of two 'moral spheres': one relating to intrahuman relations and the other concerning human–nature interactions. What needs to be ascertained is the relationship between these two distinct but related spheres of moral action. Establishing this would go some way towards bringing out the composite moral basis upon which green political theory prescribes how human social life ought to be organized. While the novelty of green political theory may lie in its concern with social–environmental issues, as a normative political theory it is social relations which are primary and from which the character of the former can be determined. Thus the critique of deep ecology in the chapter suggests that the gap between green philosophy and green politics can be overcome by focusing on the composite moral basis of green politics.

Overview of Deep Ecology

In his critical overview Dobson (1989) describes the development of deep ecology in terms of two 'turns'. The first turn refers to the fact that in its initial stages deep ecology was an environmental ethical theory concerned with the notion of the 'intrinsic value' of the non-human world (1989: 42). Its second turn was a movement away from axiology to ontology (1989: 44–6). Both are different aspects of the enduring deep ecology goal of replacing anthropocentric moral reasoning with an ecocentric moral sensibility. It is worth noting that in its second turn, deep ecology not only reaffirmed and deepened its critique of anthropocentrism, but also broke with those environmental philosophers attempting to develop an environmental ethics based on the intrinsic value of nature.

The basic aims of deep ecology are spelt out in greater detail in its 'eight-point platform'.

The eight-point platform of deep ecology

1 The well-being and flourishing of non-human life have intrinsic value, independent of human usefulness.
2 The richness and diversity of life contribute to the realization of these values and are values in themselves.
3 Humans have no right to reduce this diversity except to satisfy vital needs.
4 The flourishing of human life and culture is compatible with a substantial decrease in the human population, while the flourishing of non-human life requires this decrease.
5 Present human interference in the world is excessive, and the situation is worsening.
6 Policies affecting basic economic, technological and ideological structures must change.
7 The ideological change is mainly that of appreciating *life quality* (dwelling in situations of inherent value) rather than adhering to an increasingly higher standard of living.
8 Those who subscribe to the above have an obligation to implement the necessary changes (Devall and Sessions, 1985: 70).

The general goals of deep ecology can be stated as the preservation of nature 'wild and free' and the limiting of the human impact on nature as the way to achieve this. Breaking this down we can group deep ecology proposals under three broad headings: (a) wilderness preservation, (b) human population control and (c) simple living. These concerns are echoed by others seeking to defend an 'ecocentric' green politics, and can be taken as the hallmark of deep ecology.

This ontological shift away from environmental ethics within deep ecology has been described by Naess, the founder of deep ecology. According to him, 'The attempt to shift the *primary* focus of environmental philosophical concern from ethics to ontology clearly constitutes a fundamental or revolutionary challenge to normal environmental philosophy. It is (and should be) deep ecology's guiding star' (Naess, 1984: 204, emphasis in original). Environmental ethics *qua* axiology was viewed as the search for a convincing theory of the intrinsic value of nature. While worthy in its own way, deep ecologists came to feel it was not radical enough or sufficient to effect the types of change they thought were necessary. Leading deep ecologists argued that it was more effective to work on the way people conceive of their identity and their understanding of themselves in the greater scheme of things. For example, Sessions holds that, 'The search then, as I understand it, is not for environmental ethics but ecological consciousness' (in Fox, 1990: 225) or 'cosmological consciousness' (1990: 255).

A central part in this shift from axiology and environmental ethics to wider issues concerning consciousness reflects deep ecology's contention that there is an 'ecological crisis' which is, at root, a crisis of self-understanding and culture. That is, it is an ontological-cum-metaphysical crisis. One of the reasons given as to why deep ecology is deeper than other green moral positions is that it claims to deal with the root causes of the crisis rather than its effects. The root causes for deep ecology are found in the dominant and uninformed moral ontology of the self and a related anthropocentric culture which sees the world as dead, valueless and simply there for human enjoyment and consumption. From this ecocentric perspective the ecological crisis is first and foremost a crisis of culture and self (Eckersley, 1992a: 29). References to the need for a cultural 'paradigm shift' (Capra, 1983), based on alternative worldviews which affirm the unity of humans with and dependence upon nature, are part and parcel of the deep ecology claim that only a widespread change in consciousness will solve the ecological crisis.

Part of this shift from axiology to ontology is due to the problem of motivation or a perceived moral 'implementation deficit' in respect of theories of the intrinsic value of nature. Following O'Neill (1993), one can ask in what way the intrinsic value in nature compels us to act in a certain manner towards it. According to him:

> while it is the case that natural entities have intrinsic value in the strongest sense of the term, i.e. in the sense of value that exists independent of human valuations, such value does not entail any obligations on the part of human beings. (1993: 8)

In this way the 'ontological turn' can be seen as deep ecology's attempt to couple motivation and 'right action' by basing the latter on ecological consciousness rather than the moral discourse of the intrinsic value of nature. It is to appeal to hearts rather than minds, as it were.

Various other reasons can be found for this shift to ontological questions. For some deep ecologists the reason for this lack of attention to 'normal' ethical theorizing is that the crisis we face is too severe and deep. For Devall 'Our ontological crisis is so severe that we cannot wait for the perfect intellectual theory to provide us with the answers. We need earth-bonding experiences' (1988: 57). In other words, the shift from environmental ethics to ecological consciousness is partly driven by its perception of the causes and severity of the 'ecological crisis'. Deep ecology's evolution into an informing framework for 'living simply', and 'walking lighter on the earth', can be seen as a central aspect of its shift from environmental ethics to ontology, at the levels of both an alternative understanding of the self and an ecological way of 'being in the world' (Zimmerman, 1993). Another reason for this lack of concern with environmental ethics is the non-academic nature of much of deep ecology writing and concerns (McLaughlin, 1995). For many, deep ecology is primarily activist-orientated as exhibited in the relationship between it and the radical environmentalist group Earth First! Perhaps a more telling explanation of the shift concerns the common perception within deep ecology that 'ethics', including environmental ethics, is understood as primarily concerned with moral prohibitions, duties and obligations (Naess, 1989; Fox, 1990). For deep ecologists, duty is equated with sacrifice which is understood as the opposite to self-interest and action based on inclination.

Thus deep ecology as it presently stands concerns itself with the articulation of an alternative ontology of the self, and the place of the self in the order of nature as given by its alternative cosmology (discussed in the next section). A clear example of this is Mathews's contention that 'Deep ecology is concerned with the metaphysics of nature, and of the relation of self to nature. It sets up ecology as a model for the basic metaphysical structure of the world' (in Fox, 1990: 236). It is to the metaphysics of deep ecology that we turn next.

Deep Ecology as Metaphysics

In many respects it is unsurprising that deep ecology often comes across as a metaphysical theory, given its concerns with shifting paradigms and its tying of its critique of anthropocentrism closely to the historical emergence of particular forms of rationality, knowledge and practices in the West. The latter refer generally to the change in human–nature relations as a result of the Enlightenment. Thus, deep ecology's critique of anthropocentrism is sometimes an expression of its more general critique of 'modernity' (Oelschlaeger, 1991; 1993; Zimmerman, 1993). Within deep ecology therefore it is often difficult to separate out the critique of anthropocentric moral reasoning from this different, and in many ways more contentious, critique of modernity. This equation of modernity with anthropocentrism can be readily seen in deep ecology's standard historical account of the 'disenchantment of nature': the transformation of nature from a realm of meaningful normative significance into a collection of resources for human instrumental use and exploitation (Fox, 1990; Barry, 1993). The historical shift to a mechanistic, reductionist, instrumentalist worldview is what deep ecology means by 'anthropocentrism', as suggested later. Anthropocentrism thus refers to a complete metaphysical worldview, one deep ecologists claim as the root of the ecological crisis. This worldview is held to underpin all dominant moral theories and political ideologies, apart from the deep ecological one. From this deep ecological viewpoint 'the Green movement . . . is self-consciously seeking to call into question an entire world view rather than tinker with one that already exists' (Dobson, 1990: 8). Hence the necessity for a 'new metaphysics' to underpin human–nature relations, premised on transcending anthropocentrism and replacing it with 'ecocentrism'. However, as will be argued below, this rules out the *politically* powerful strategy of basing green politics on the *immanent critique* and reformulation of anthropocentrism (and by implication 'modernity'). Thus while a reformulated anthropocentrism is more defensible philosophically (as argued in the next chapter), it is, *a fortiori*, more defensible *politically*.

Deep Ecology and the 'Re-enchantment of Nature'

This section looks at some salient aspects of deep ecology metaphysics understood as the re-enchantment of nature. There are three main issues that this re-enchantment theme highlights. The first is the role of re-enchantment within deep ecology's critique of modernity, based on its diagnosis of the 'ecological crisis'. The second is the spiritual complexion this concern with re-enchanting nature lends to deep ecology. The third issue relates to deep ecology as a 'politics of redemption'.

One way to understand deep ecology can be found in a seminal essay by Lynn White on 'The Historic Roots of our Ecologic Crisis' (1967). He concluded that 'Since the roots of our trouble are so largely religious, the remedy must also be essentially religious, whether we call it that or not. We must rethink and re-feel our nature and destiny' (1967: 1207). This demand for a metaphysical 'paradigm shift' is an enduring feature of deep ecology (Capra, 1983; 1995; Naess, 1989: 20; Mathews, 1991: 40–1), which can be seen as a critical reaction to the 'disenchantment' of nature, and the 'dominant paradigm' or worldview which caused this disenchantment. Deep ecology follows Horkheimer and Adorno's argument that 'the program of the Enlightenment was the disenchantment of the world, the dissolution of myths and the substitution of knowledge for fancy' (1973: 3). In short, science and technology, as the quintessentially 'modern' and dominant 'rational' ways of understanding and interacting with the world, have disenchanted and desacralized nature. In setting its face against this disenchantment, deep ecology rejects the idea that it was either an inevitable or a worthwhile price to be paid for the 'benefits' of modernity. The thrust of deep ecology follows the logic of White's argument suggesting that if the cause of the ecological crisis is to be found in the disenchantment of nature then the solution lies in its re-enchantment.

A common interpretation of deep ecology is that its solution to the ecological crisis lies in a widespread quasi-religious conversion along the lines suggested by Clark for whom 'Only a "religious spirit", a willed and eager commitment to a larger whole, can easily sustain us through adversity let alone through prosperity' (1994: 114). Seen from this general framework, deep ecology is a spiritual/religious answer to the ecological dilemmas facing humanity as a result of modernity and its attendant anthropocentric worldview. If this is the case then deep ecology renders green politics as spirituality by other means; that is, concerned not with coping with the ensemble of ecological problems facing us, but with discovering the 'truth' of our metaphysical relation to the non-human world as the normative basis for resolving the 'ecological crisis'. Rather than elaborating the context within which agreement on the normative rightness of social–environmental affairs can be created, deep ecology appears to seek a determinate answer to the existential riddle of the 'human condition'. The importance of this spiritual self-understanding within deep ecology is that this has been taken to excuse its lack of a political dimension. For example, according to Devall, 'The deep, long-range ecology movement . . . is only partly political. It is primarily a spiritual-religious movement' (1988: 160). The 'ecological crisis' for deep ecology is thus a 'crisis of civilization', rather than a potentially resolvable contradiction within present cultures. The reconstructive aim of rethinking green politics is to see that talking of the 'ecological crisis' (which has been the dominant claim around which it has presented its ideas) may be damaging for its political effectiveness, and not just its descriptive accuracy of the problem facing contemporary societies.

One way of highlighting this aspect of deep ecology is to ask: what does the resolution of the ecological crisis mean for it? Within its terms of reference, it seems that only a final solving of the problem between humanity and the world constitutes a solution. Solutions which prevent the destruction of the natural world but are not motivated by a reverence for nature fall short of the deep ecology standard. This is clear from Seed's contention that 'Deep ecology recognizes that nothing short of a total revolution in consciousness will be of lasting use in preserving the life-support systems of our planet' (1988: 9). To paraphrase Dobson, for deep ecology the reasons for care for the world are as important as the care itself (1989: 46). Care motivated by anthropocentric reasons falls short of the deep ecology ideal. However, this 'solution' goes well beyond that needed to function as a normative agreement for the achievement of green social and political values and environmental policies. In conceiving of the problem in this light, deep ecology prioritizes discovering some putative metaphysical *truth* of the 'human condition', over the political task of constructing a public and normative basis for *social co-operation* to achieve ecological sustainability within a moral framework for human/non-human affairs.

One may say that deep ecology's concern with finding the 'truth' of human–nature relations leads it to seek a *permanent* solution to human–nature relations, in the sense of finding the definitive, final, once-for-all answer to this aspect of the human condition. In this way, there is more than a hint of 'closure' within the deep ecology position. The search for permanent solutions to human–nature relations may also account for the frequency with which a return to the values (if not the practices) of hunter-gatherers can be found in some deep ecology writing (Shepard, 1993). This is because these modes of human life existed in an almost permanent equilibrium or harmony with nature, largely because they simply had to accommodate themselves to the natural order to survive. Thus the arguments for wilderness protection, population control and simple living, mentioned above as litmus tests for deep ecology, can be explained, at least in part, by a sense that hunter-gatherer society represents the 'ideal' form of human society which we moderns must emulate.

Against the deep ecology notion of finding a permanent solution, green politics, I argue, should accept that social–environmental relations will always be characterized by uncertainty, contingency and possible catastrophe, partly as a result of human transformative interests and practices, partly because of the limits to human knowledge of the world used in fulfilling those interests, and partly because nature itself poses threats to our existence through 'natural disasters'.

One implication of deep ecology's search for a permanent solution is that normative agreement sought by deep ecology goes beyond a shared *respect* for nature to a much stronger argument concerning the necessity for a shared *reverence* for nature. That deep ecology is more concerned with reverence than respect can be seen in Mathews's statement that: 'When our culturally-endorsed cosmology represents the world as inert, blind, bereft of worth or purpose, indifferent to our attitudes towards it, *then our natural urge to celebrate Nature may be thwarted*' (1991: 162–3, emphasis added). This idea of a 'natural urge to celebrate Nature' is also tied up with deep ecology's diagnosis of the 'pathology' of the modern self which becomes 'ill' when such 'natural' urges are repressed (Kidner, 1994). The deep ecology aim to re-enchant the world is understood as a return to a 'natural' harmony between humans and nature. It would not be out of place to claim that clarifying deep ecology along these lines highlights its character as a 'politics of redemption'. As a redemptive politics deep ecology can be seen as a critique of modernity which views the Enlightenment as humanity's 'second fall'. It is interesting to note that, as with the original fall, deep ecology's position is that the ecological crisis is a result of human arrogance, premised on the possession of knowledge. This would certainly tie in with the view that the ecological crisis is nothing short of a total crisis for humanity, and it would also go some way to explaining the spiritual overtones of much deep ecology as well as its anti-urban pastoralism.

Finding normative agreement for saving the planet, however, does not demand that green politics tie this to saving souls or finding the answer to the meaning of human life written in the text of nature, or that a permanent solution to social–environmental relations can be equally 'read off' from nature. A final interpretation of the re-enchantment theme is that it refers to the resolution of the alienation of humans from nature, that is the result of a worldview which stresses the separation of humans from nature. Re-enchantment is here related to a therapeutic, healing concern for the damaged or immature modern self, which is alienated from a redemptive contact with nature.

The ontological reorientation of deep ecology is thus largely to do with a qualitatively different metaphysical understanding of the world (and the cosmos), and the place of humans within it. Despite the claims of Naess and others who hold that there are many different metaphysical and cosmological 'ultimate premises' that are compatible with the 'deep ecology platform' (Naess, 1988: 129; McLaughlin, 1995), it is clear that deep ecology is committed to 'enchanting' rather than disenchanting cosmologies (Mathews, 1991), of which there could be many. A good example of this is Fox's claim that 'deep ecology recognizes that an ecologically effective ethics can only arise within the context of a more persuasive and more enchanting cosmology than that of mechanistic materialism' (1984: 195).

Deep Ecology and Psychological Health

Consonant with its general drift, for deep ecology the ecological crisis is tied up with Western culture's pathological understanding of the self as a result of its disenchanting, anthropocentric worldview (Naess, 1989; Fox, 1990; Kidner, 1994). For deep ecology the ecological crisis is due, in part, to a misconceived notion of the 'self', which unnecessarily limits the scope of moral concern and care. The anthropocentrism inherent within dominant Western conceptions of the self is held to set up a subject/object dualism which both reduces subjecthood to a narrowly conceived human self, and reduces non-human nature to the status of pure object. For Naess (1989), a reconceptualization of the human self is a necessary precondition for a less anthropocentric perspective. In psychological terms, the modern, unecological self is viewed as immature, sick or underdeveloped. Warwick Fox goes much further in his criticism and asserts: 'Anthropocentrism represents not only a deluded but also a dangerous orientation toward the world' (1990: 13) – dangerous in terms of the health of the self as well as the health of the planet.

The psychological turn of deep ecology can be expressed in the following schema:

1 The 'self' within Western moral thinking is narrowly conceived: it is not false but incomplete/immature/unhealthy.
2 This view of the self expresses an anthropocentric view of the world and our place in it, and is one of the main causes of the ecological crisis.
3 What is needed is a more expansive notion of the 'self', an understanding that is premised on the *inclusion* as opposed to the *exclusion* of the non-human world.
4 This ecocentric 'sense of self' can be cultivated by a process of 'identification' with the non-human world.
5 This sense of self involves an orientation of care to the non-human world.
6 Injunctions, rules and principles, i.e. deontological ethics, become superfluous to ensure morally appropriate human interaction with the non-human world.

The Kantian distinction between 'beautiful' and 'moral' actions is used by Naess (1989) to explain the consequences once the 'ecological' self has been realized. For him, this means that

> Moral actions are motivated by acceptance of a moral law, and manifest themselves clearly when acting against inclination. A person acts beautifully when acting benevolently from inclination. Environment is then not felt to be something strange or hostile which we must unfortunately adapt ourselves to, but something valuable which we are *inclined* to treat with joy and respect. (1989: 85)

The solution of the moral dilemma does not consist in the formulation of 'environmental ethics' (understood as a system of moral 'oughts' coming from an environmentally informed axiology). Although acknowledged as having a role to play, the latter is regarded as inferior to encouraging ecocentric habits and dispositions so that 'beautiful' ecological actions follow 'naturally'. This process of 'ecological identification' (point 4 in the schema given) is part of the 'maturing' of the self according to Naess (1989: 86). It is the 'immaturity' of the self and the type of ethics associated with that view of the self which result in a disvaluing of nature and a simultaneous underestimation of the individual. In treating nature disrespectfully we simply reveal our misunderstanding of what it means to be a 'mature' self. We treat nature as we do because we have 'forgotten' our place in nature, 'we do not understand who we are' (Dodson-Gray, 1981: 84).[1]

This psychological theme and concern with the ontology of the self will be looked at through an examination of the work of Naess and Fox. For Fox, the central ideas of deep ecology are to be understood within a psychological framework. His 'transpersonal ecology' is an extension of Abraham Maslow's work on 'transpersonal psychology'. (Indeed,

Maslow's theory of a 'hierarchy of needs' which culminates in 'self-realization' is another key aspect of the psychological perspective of deep ecology: Naess, 1989; 1995; Fox, 1990: appendix B.) For Fox

> the response of being inclined to care for the unfolding of the world in all its aspects follows 'naturally' – not as a *logical* consequence but as a *psychological* consequence; as an expression of the spontaneous unfolding (developing, maturing) of the self. (1990: 247)

Echoing Naess, he states that this view of the self has the

> highly interesting, even startling, consequence that ethics (conceived as being concerned with moral 'oughts') is rendered superfluous! The reason for this is that if one has a wide, expansive, field-like sense of self then . . . one will naturally (i.e. spontaneously) protect the natural (spontaneous) unfolding of the expansive self (the ecosphere, the cosmos) in all its aspects. (1990: 217)

Basically, ecological selves do not harm their environment because the process of identification with it highlights their continuity with the environment. This sets up an expanded conception of the self such that the environment is constitutive of their ecological sense of self. To diminish or degrade the environment with which they identify would be to diminish or degrade themselves. Consonant with an expanded notion of self, there is an equally expanded notion of self-interest such that acting in one's self-interest from desire and inclination will not result in the disrespectful treatment of nature. Behind the stress on 'identification' and the ecological self is a belief that 'natural' self-interest can be suitably altered so that positive ecological benefits result from this ecological way of 'being in the world'.

However, there is a potentially unresolvable tension within this approach, that between means (beautiful actions of the ecological self) and ends (the preservation and protection of wilderness for example). Behind the stress on 'beautiful action' there is an objectivist assumption that all will think and act in the same way. We can ask why beautiful action in regard to the environment is not simply in the eye of the beholder.[2] Extending selfhood to include the non-human world, and encouraging action based on inclination and desire rather than self-sacrifice and obeying the 'moral law', are equally compatible with the preservation of nature or its development depending on how the 'ecological self' views its self-interest. What guarantees that the ecological self will absorb the needs of nature, and thereby incorporate them as its own, is the deep ecology metaphysic, which is unlikely to command widespread support. One can conclude by saying that focusing on the 'ecological self' is insufficient to understand the deep ecology position. Ultimately it is its metaphysics which is the final arbiter of what constitutes 'beautiful actions'. Re-enchantment of nature goes hand in hand with psychological reconnection and the overcoming of the self's alienated state. It is the metaphysical vision of deep ecology which furnishes the objective criteria by which ecological selves are to be judged.

Knowledge, Ontology and Metaphysics

Deep ecology's general frame of reference is the restoration of something that has been lost, a return to the true path from which we have diverged. Aboriginal cultures are often held up as exemplars of good ecological behaviour by deep ecologists. 'Wilderness experience' is regarded as a close approximation to these ancient and 'true' human ways of 'being in the world'. This can explain much about deep ecology. For one, it explains the stress placed on the direct experience of nature, rather than reasoning about nature. Via the direct apprehension of the natural world, either by living close to nature or by 'wilderness experience', we may relearn and recover our place in the natural order of things, rediscover the perennial rhythms of the earth, and once again be in harmony with the world (Kohák, 1984; LaChapelle, 1993).

The use of terms such as 'health', 'well-being' and 'maturity' within deep ecology writing indicates that there is an 'objective' standard against which we can assess the development of the self. For example, the emphasis on the 'maturing of the self' implies that Fox (1990) has some conception of the self and its 'proper' development. From the deep ecology point of view those who do not adopt its position (or rather have *yet* to adopt it) are 'immature' or 'undeveloped'. Another interpretation, and one more consonant with the 'psychological' component of deep ecology's 'ontological turn', is the idea that for deep ecology the ecological crisis is an expression of a deep-seated 'pathology' that must be overcome 'if we are to develop a healthier relation to the natural world' (Kidner, 1994: 45). Within this 'therapeutic discourse', the resolution of the ecological crisis is intimately tied up with the simultaneous resolution of the 'psychological crisis' induced by the 'modern worldview' or 'dominant paradigm'. This is of course related to the deep ecological view of the ecological crisis as a 'total crisis' – a constitutive part of which is a crisis of self-understanding.

A serious problem with this 'psychological' turn of deep ecology is the objectivism and essentialism associated with its theory of the self. What deep ecology attempts to provide is a classificatory schema by which to judge and rank self-understandings. Self-understandings that depart from its standards are misconceived, incomplete or 'sick'. My point is not that there are no standards by which we can judge understandings of the self, but rather that such standards are not 'given' or 'discovered' in some objective fashion to be 'read off' from some metaphysical-cum-spiritual schema. Rather, they are the product of an ongoing process of intersubjective negotiation and discourse, premised on the shared activities of human beings.

By construing the issue as one of 'moral health and development', the implication is that there are discernible qualities that 'healthy' and 'mature' selves exhibit. The paradigm case of this is 'wilderness experience': a transformative act which simultaneously 'reveals', in a manner that is not always clear, a mode of human experience and moral reasoning 'superior' to anthropocentrism. Communing with nature by working one's allotment does not carry the same transformative power. 'Unhumanized' nature, i.e. wilderness, alone has this capacity. Selves which do not identify with the non-human world are alienated, unhappy selves. Thus, deep ecology's view of the 'unhealthy self' fits with its view of modernity as the 'disenchantment of nature', and Fox's (1990) view of anthropocentrism as a 'dangerous orientation to the world'.

Deep ecology, in articulating its proposals within a psychological theory of moral development, reformulates the normative 'ought' with respect to nature to become a matter of discerning what 'is' in the enlightened interest of the ecologically expanded self. And this is discovered metaphysically in an *a priori* fashion, rather than as an outcome of intersubjective discourse. The displacement of 'moral action' by 'beautiful action' could never be anything more than temporary, given the irreducibly plural (and thus conflict-prone) nature of the 'moral sphere' as a generic space of human action, and mode of being.[3] Within deep ecology we '(re)discover' the 'ecological self' rather than intersubjectively 'create' it through a shared ethical discourse. It is discovered by reference to the metaphysic of oneness and/or the practices and belief systems of surviving human societies which express the deep ecology metaphysical sensibility. A serious problem with this mode of reasoning is the understanding and critique of anthropocentrism associated with it.

Sensitivity to the various gradations within anthropocentrism is blunted by the definition of anthropocentrism used by deep ecologists. Eckersley defines it as

> the belief that there is a clear and morally relevant dividing line between humankind and the rest of nature, that humankind is the only or principal source of value and meaning in the world, *and that non-human nature is there for no other purpose but to serve humankind.* (1992a: 51, emphasis added)

Breaking this statement down into three propositions, we can discern different aspects of the deep ecological critique of anthropocentrism.

Firstly, that there is a morally relevant divide between humans and non-humans is a statement that all except committed biospheric egalitarians would agree with. As explained in the next chapter, being human counts for something in a way which the charge that anthropocentrism is simply 'ungrounded speciesism' fails to register (Routley and Routley, 1979). That the difference between humans and non-humans may be one of degree rather than kind does not deflate the importance of this basic distinction between how humans interact with each other and how they interact with the rest of the world. As will be recalled, in the introduction I argued that the moral basis of green political theory is a composite one, made up of two moral spheres, one human only and the other concerning social–environmental relations.

The second statement, that humans are the only morally relevant beings in the world, does not follow from the first. Accepting our status as the only or main source of value and meaning in the world can ground widely different attitudes to the world. From this perspective anything from the complete and unhindered exploitation of the world (the third statement) to the widespread protection of vast tracts of nature from human interference can be forthcoming. For purely human reasons, informed by the idea that we attribute value in an otherwise valueless world, our action in the world can be either extensive or minimal, and is compatible with extending moral considerations to human interaction with the non-human world. And finally, there is nothing inherently ecologically unfriendly about the fact that humans, as far as we know, are the only species with a moral sense.

It is the third statement that goes to the heart of the deep ecology position, where anthropocentrism is understood as expressing an exclusively strong instrumentalist conception of the world. In Eckersley's formulation the non-human is said to be exclusively of instrumental concern to the anthropocentrist. However, it does not follow, either logically or in practice, that the first two positions lead to this instrumentalist one. This hypothesis can be viewed as, in part, a metaphysical claim, the idea that the world exists solely for human use and enjoyment. Without such a metaphysical context it is difficult to see how this statement can be meaningful. Unlike the previous two arguments, which can be thought of as expressing general ethical features, the presumption that the non-human world is there purely for human use is a meta-ethical claim. The problem with anthropocentrism, for deep ecology, is thus not how it operates as an ethical theory, since ethics is meaningless outside an anthropocentric context, but how it operates as a metaphysical position about the place it accords humans within the natural order. Yet there is no necessary reason why this strange claim – which, religious justifications apart, would be extremely difficult to establish – should be thought of as *essential* to anthropocentrism. The real reason why deep ecologists are suspicious of reformed anthropocentrism is that the status of the non-human world remains contingent: that is, *it does not enjoy permanently protected status*. This suspicion rests on the fact that deep ecology has at its heart an *a priori* position which privileges the *preservation* of nature over the human use of nature (see below).

The Normative Basis for Political Agreement

A strong argument against deep ecology is that it is much easier to secure agreement about the normative rightness of an action or practice at the level of applied ethics than at the metaphysical or philosophical level, which is the main focus of deep ecology. At the applied ethics level, agents may have different reasons for agreeing to the same policies. And ultimately we must ask which is more important, achieving social agreement around green policies and practices, or seeking consensus on the reasons for those changes? Dryzek (1987) notes that environmental issues are such that it is often the case that agreement on the policies for pollution abatement or sustainability can be achieved, despite different reasons that may be advanced for those policies. In a later work he argues that 'disagreement on the fundamental principles of morality (pure ethics) often proves compatible with consensus on the moral side of practical issues (applied ethics). In other words, *prudential, context-specific* moral reasoning can overcome differences in abstract commitments' (1990: 17, emphasis added). Agreement around *what* should be done can be independent from *why* it should be done, in the sense that various reasons can be given for the same or similar action. This is also echoed in Norton's (1991) 'convergence hypothesis', discussed in more detail in Chapter 5. According to him, the fact that greens (of whatever hue) increasingly agree on policy while disagreeing on ultimate principles is a positive advancement, and one that can be used to overcome the ecocentric–anthropocentric dichotomy. That is, green arguments and policy proposals would receive a better hearing by the public if environmental policies were cast in terms of extended human interests, rather than emphasizing non-human interests. A clear example of this is environmental policy based on a moral concern for future generations.

However, deep ecology's demand that 'the reasons for the care of the non-human world are at least as important as the care itself' (Dobson, 1989: 46) greatly diminishes the probability of deep ecology securing sufficient normative agreement publicly for green principles. The question that needs to be asked is: can deep ecology function as the shared political basis of normative agreement between citizens upon whom such policies will be binding? Will deep ecology secure public agreement for green policy prescriptions? Given the major changes that the resolution of ecological problems may require and the consequent disruption of citizens' lives, the issue of securing agreement for such change is obviously crucial.

Employing Rawls's distinction between 'political' and 'metaphysical' bases of political morality, we find that deep ecology is 'metaphysical' not 'political', and as such cannot underwrite a political agreement around green political ends. According to Rawls, 'as a practical political matter no general moral conception can provide a publicly recognized basis for a conception of justice in a modern democratic state' (1985: 225). But this is precisely what deep ecology attempts to do. It seeks to secure agreement and support for principles and policies on the basis of its particular philosophical understanding of the self, the non-human world and the relation between the two. And what is more it states that social restructuring aimed at securing ecological sustainability will fail if not based on its prescriptions. It will 'fail' in the sense that political prescriptions and policies that have not come out of a process of 'metaphysical reconstruction' will be 'unworthy', or fall short of deep ecology ideals, even if successful by other criteria, such as ecological sustainability. Deep ecological arguments for altering the basic institutions of society will not succeed in gaining sufficient normative support. The point is that it is unlikely that deep ecology will succeed as a stable basis for normative agreement.

The problem with deep ecology is that it brings green politics into irresolvable conflict with settled convictions, giving it a 'fundamentalist' complexion which is a hindrance to convincing non-believers to support its political aims. This does not imply that green politics must simply accept the prevailing anthropocentrism as 'given' and beyond change. Far from it. *The defining feature of green moral theory should be not the acceptance of ecocentrism but a critical attitude to anthropocentrism*. However, the process by which such change is secured is a political matter, doing moral theory in the public sphere as it were, and accepting that persuasion must start from some degree of underlying agreement. That such activity is human-centred does not mean that it is solely concerned with human welfare or interests. As argued in the next chapter, an acceptance of the appropriateness of an anthropocentric 'moral extensionism' is a more secure basis upon which to argue for green politics, not only because it is conceptually a more defensible moral position, but because politically it seeks to persuade within a generally accepted discourse. Its strength lies in its *immanent* character as opposed to the *external* nature of non-anthropocentrism. If the public justification of green politics is derived from deep ecology, it is unlikely to be accepted by a majority of the audience to whom it is addressed.

If deep ecology is correct and the ecological crisis cannot be resolved without a reconceptualization of the self along the lines of the 'ecological self' within a more widespread change in cultural self-understanding, then green politics is in trouble (or, from the deep ecology perspective, superfluous or not a priority). This is because as understood here, green political theory *does not* see the ecological crisis as fundamentally an 'ego-crisis of humanity', as opposed to presenting a constellation of moral and material contradictions which can be resolved *within* the present culture and its forms of normative reasoning. Rather than demanding a 'change in the context in which the [ethical rule] book is written' (Dobson, 1989: 44), I suggest that an *immanent critique* of conventional (i.e. anthropocentric) moral reasoning is sufficient to establish the normative claims of green politics. This is a basic difference between deep ecology and other environmental theories. Deep ecology is premised on viewing the ecological crisis as a 'total crisis', a fundamental 'crisis of civilization' (Bahro, 1994). From this perspective, it is only by transcending anthropocentrism (as a form of moral reasoning and culture) that this 'total crisis' can be resolved. Ethical theory is accordingly viewed as so embedded within the prevailing 'modernist' or 'industrial' anthropocentric culture that it simply cannot escape and formulate sufficient alternative ethical guidelines and modes of interaction. McLaughlin (1994), in a chapter tellingly entitled 'Beyond Ethics to Deep Ecology', highlights this aspect of deep ecology. According to him:

> The social dependency of ethical theory is a serious problem for any attempt to develop a non-anthropocentric environmental ethic. If the issues posed by ecological crises go to the very roots of industrial society, *then it is unlikely that any ethical theory that is grounded in reflection on current social practice will penetrate deeply enough. . . . Thus, the possibility of grounding ethical argument for any radical transformation of humanity's relations with the rest of nature requires going far beyond ordinary ethical discourse.* (1994: 169, emphasis added)

McLaughlin's claim that an immanent critique of anthropocentrism and its cultural manifestation 'will not penetrate deeply enough' is premised on a presumption that the only solution must be an ecocentric one. Alternatives, both from non-anthropocentric environmental ethics, and more importantly, from within anthropocentrism, are dismissed. This, as suggested in the next chapter, is to throw the baby out with the bathwater. The argument that ethical reflection is constrained by contemporary social practices is both misleading and self-serving. That conventional ethical theory may be grounded in reflection on those social practices does not prove that the former is determined by the latter or that it cannot radically change practices. One has only to survey the recent history of moral and legal theory to see the effect it has had on social practices from women's rights to the legal protection of some non-humans, including most famously trees having 'standing' in the US (Stone, 1974). In common with radical theories, deep ecology is simply too *impatient* to work within the conventional moral discourse on human–nature relations. Thus its ontological turn is motivated partly by a desire for rapid cultural change, without recognizing the central place of democratic politics in that process of transformation.

It is more likely that agreement will be reached on an 'ethics of use' for social–environmental relations rather than on a metaphysical truth of the place of humans in the wider natural order. In this sense the early Rawls leads us in the wrong direction when he claims:

> A correct conception of our relations to animals and to nature would seem to depend upon a theory of the natural order and our place in it. One of the tasks of metaphysics is to work out a view of the world suited for this purpose; it should identify and systematize the truths decisive for these questions. (1972: 512)

From the point of view of green politics, Rawls can be criticized for viewing social–environmental relations as a metaphysical issue, which seems to rule out agreement on environmental matters as part of a wider normative agreement on social co-operation, given the quote from him earlier. The green argument is that these relations can be viewed as a normative issue upon which it is possible to find some degree of political agreement at the level of applied ethics and policy. In conclusion, green politics ought to reject the conversion of large sections of the population to a deep ecological worldview as a key political aim. However, if some level of metaphysical agreement is needed, then science may offer a sufficient basis for such agreement.

Science as a Basis for Metaphysical Agreement

If green politics is to base itself upon some metaphysical footing, then science rather than an earth-centred spirituality may be a much better way of going about it. As Grey notes, 'A purely secular, scientific naturalism can provide a thoroughly satisfying way of realizing our unity with the non-human world' (1986: 212). Thus modern science, a product of modernity it needs to be noted, can offer an account of the place of humans in the natural order. That is, modern forms of scientific knowledge can help displace the arrogance of humanism, without rejecting anthropocentrism. There is no basis in science for holding that the earth and all it contains was created for the human species and it alone. Thus science can dispel any potential arrogance of anthropocentrism. One way in which science, including ecological science, can do this is by demonstrating that the relationship between 'human' and 'non-human' is characterized by both *difference* and *differentiation*. That is, science can promote a worldview in which the human condition is marked by being *a part of but also apart from* the natural order (Barry, 1995c), undermining any notion of a radical or unqualified separation of humans and nature. This scientifically informed metaphysical view of human–nature relations, premised on the idea that the world is our home, but one which we share with a variety of other living and non-living entities, and with which we are connected in important ways, can, I suggest, function as a sufficient metaphysical basis for green politics.

Norton in his plea for unity among environmentalists demonstrates the significance of science within green politics. According to him: 'Environmentalists' emerging consensus, it will turn out, is based more on scientific principles than on shared metaphysical and moral axioms' (1991: 92). He then goes on to point out that:

> The attack on human arrogance, which was mounted as a response to anthropocentrism, was well motivated but badly directed. One need not posit interests contrary to human ones in order to recognize our finitude. If the target is arrogance, a scientifically informed contextualism that sees us as one animal species existing derivatively, even parasitically, as part of a larger, awesomely wonderful whole should cut us down to size. (1991: 237)

Within contemporary Western society it is more likely that a non-spiritualized scientific understanding of the world and our species' place in it can provide basic metaphysical agreement. Scientific knowledge can reform anthropocentrism, by lessening its tendencies to hubris and pride. That is, it can help avoid the ecological vices of the latter. It can also demonstrate our dependence upon the environment, which as will be argued later is vital for underpinning the virtues of ecological stewardship. It is also obvious, as O'Neill points out, that 'Scientific theory and evidence are a necessary condition for a rational ecological policy' (1993: 145). By this he means that scientific verification of ecological problems will be a necessary, though not sufficient, condition for social recognition that there are problems. At the same time agreement on the scientific nature of ecological problems can be useful in forging a politically workable normative agreement on social–environmental issues. The place of scientific knowledge within green moral and political theory will be further developed in the next chapter and Chapters 5 and 7. All I wish to note here is that a possible metaphysical basis for green political theory can be found within a secular scientific naturalism.

In the end, for deep ecology, an 'ethical' articulation of human relations to nature is in some way to admit falling short of being in harmony with it. An environmental ethic signifies a lack within human–nature relations of compassion, of sensitivity on behalf of humans. In this respect, the deep ecology critique of environmental ethics is similar to the communitarian critique of justice. From the communitarian position, the appeal to justice as the normative basis of social co-operation signifies a lack of solidarity and fellow-feeling within society as a whole. On this account, justice is, at best, a 'remedial virtue' (Kymlicka, 1993). In a similar communitarian fashion, deep ecology criticizes environmental ethics on the grounds that it is an inferior or second-best normative basis for regulating social–environmental relations. This can be seen in Naess's comment that, 'We need not say today man's relation to the non-human world is *immoral*. It is enough to say that it lacks generosity, fortitude, and love' (in Fox, 1990: 221). It is for this reason that identification is so important within deep ecology since, as Naess again points out, 'there must be identification in order for there to be compassion [between humans and non-humans], and amongst humans, solidarity' (1995: 227). The ethical is unnecessary, and may indeed be unhelpful, while the political is something which can be derived from discovering our place in the sun and the order of nature. As Eckersley states, 'In terms of fundamental priorities, an ecocentric approach regards the question of our proper place in the rest of nature as logically prior to the question of what are the most appropriate social and political arrangements for human communities' (1992a: 28). This again construes human–nature relations as primarily a metaphysical as opposed to an ethical matter, as well as privileging human–nature relations over intrahuman ones.

The development of an ethical anthropocentrism divorced from a metaphysical one opens a common space between deep and non-deep green moral theory. It is the *arrogance* of humanism (Ehrenfeld, 1978) rather than humanism itself that is, or ought to be, the proper object of the deep ecology critique, and should be a central orientating feature of green moral theory. This arrogance ought to be seen as an ecological vice to be held in check by ecological virtue. This is in keeping with Naess's claim that it was sufficient to criticize human treatment of the environment on the grounds that it lacked generosity, compassion or love. In other words, one could say that the deep ecology argument is against an uncaring, economic, narrow-minded humanism rather than against humanism itself. A humanism which 'honoured' the spirit of deep ecology would be compatible with much of what deep ecology seeks to achieve. Indeed, as outlined in the next section, deep ecology's compatibility with a version of virtue ethics offers a way in which it can be incorporated within an alternative, but still anthropocentric, basis for green politics. Virtue can act as a bridge between deep ecology and a reformulated anthropocentric ethical theory.

From Deep Ecology to 'Ecological Virtue'

For deep ecologists the 'ecological crisis' is partly due to the 'illusion' of human technological and epistemological prowess *vis-à-vis* the natural order. Crudely put, the ecosystems upon which all life is dependent are beyond human comprehension and the idea of human control over the ecological conditions of life is a dangerous presumption. According to Naess 'Such complexity [of ecosystems] makes thinking in terms of vast systems inevitable. It also makes for a keen, steady perception of the profound human ignorance of biospherical relationships and therefore of the effect of disturbances' (1973: 245). The overarching idea of deep ecology could be summed up as: that which we cannot understand we cannot control, and in regard to nature the appropriate attitude towards that which we cannot control is humility, awe and reverence. A typical example is Devall and Sessions who note that the deep ecology movement is concerned with 'cultivating the human *virtues* of modesty and humility' (1985: 110, emphasis added). This concern with moral character and habits, however, can be integrated within a reformed anthropocentric green moral theory.

In stressing the significance of identity within moral experience, deep ecology has something positive to bring to green politics in terms of a critique of the lack of 'concreteness' in 'rationalist' accounts of ethics, which I discuss in more detail in the next chapter. Interpreted as a form of virtue ethics, deep ecology suggests how care for the environment and nature can be related to the self in a way which bypasses many of the problems associated with attempts to ground care in the intrinsic value of nature. Stressing the cultivation of certain virtues as indispensable action-guiding qualities allows aspects of the 'ecological self' to be incorporated within green political theory, divested of the limitations mentioned above. Virtue ethics, in short, may offer a less contentious way in which human identity and character can be made more central to the understanding of moral experience than the metaphysical or psychological views of the ecological self.

For example, unlike the deep ecology account of the 'ecological self', a virtue-based account of how we ought to treat the non-human environment would make the social and cultural dimensions of such treatment explicit. That is, a virtue-based account of the moral treatment of the non-human world makes sense if it is informed by and consonant with some socially constituted valuation of the non-human world. Thus, a virtue-based account, unlike 'rationalist' accounts of ethics, focuses on the character of the individual agent. It highlights the particular cultural valuations of the natural world which form the background and framework within which character formation and individual moral action take place. Such accounts of ethics are therefore typified by a sensitivity to the context and particularity of actual, concrete social–environmental relations. It follows from this that a virtue ethics approach means that any attempt to reorientate the moral relationship between humans and the natural world automatically involves reference to social relations. Virtue-based accounts of our ethical treatment of the non-human world are thus *political* in a way which the 'ecological self' of deep ecology is not, while sharing the latter's concern with relating ethics to character.

Adopting a virtue ethics perspective recasts the critique of anthropocentrism. From this perspective the arrogance, hubris and inflated self-importance that can characterize the latter are *vices*: that is, unworthy (and damaging) moral attributes or dispositions of character. Thus a virtue approach can enable green moral criticism to find the correct target: the 'arrogance' rather than the 'humanism' as indicated above. Virtue ethics also allows the positive attributes of the ecological self to be understood as specific environmental virtues. For example, we can say that deep ecology suggests a degree of humility and compassion to counter the excesses, i.e. vices, of anthropocentrism.

However, if we accept the Aristotelian definition of a virtue as the mean between two vices/extremes, deep ecology can be criticized for sometimes lapsing into the opposite extreme of 'arrogant humanism', namely a complete submersion within, and total acceptance of, the order of nature. The latter, according to Frasz, is typical of 'someone who has lost all sense of individuality when confronted with the vastness and sublimity of nature' (1993: 274). This is the danger of both 'wilderness experience' and a desire to return to or recapture the values of a hunter-gatherer form of society. This opposite extreme of 'arrogant anthropocentrism' typifies many unreflective sentimental or romantic views of human–nature relations which pepper green moral and political discourse, sometimes expressing itself as 'quietism' (Wissenburg, 1993: 9). And as an extreme this moral disposition is thus not an environmental virtue, but rather an environmental vice. Sentimentality with regard to human–nature relations does not give nature or humans their proper regard since it often occludes 'negative' aspects of this relationship, such as predation, use, consumption, labour and death, which are, in reality, inescapable 'facts of life', part and parcel of the 'human condition'. We can thus, following classical moral thought, distinguish between the 'intellectual virtues' (knowledge, wisdom) and those associated with character (such as humility, generosity, friendship), what Aristotle called the 'moral virtues'. Ignorance of our dependence on nature, as given by ecological and other sciences, which tell us that we are, for example, social animals evolved from primates, and that we are a part of the natural order, is a vice, and *a fortiori* an environmental vice. Thus knowledge of the ecological facts of life is an environmental virtue to be cultivated. For humility to be the virtue deep ecology supposes it to be, it must be understood as a mean between a timid ecocentrism and an arrogant anthropocentrism. As I argue in the next chapter, weak anthropocentrism occupies this mean between these two extremes.

Ecological Virtues and the 'Human Condition'

Virtues are qualities of character enabling their possessor to be responsive to the inherently contingent and contextual character of human experience. That is, virtues, such as courage for example, are dispositions which humans need or find useful in order to live because of the type of beings they are and the type of world(s) they inhabit. It is important to point out that virtues are commonly held to help human beings to *cope* with rather than to *eliminate* the problems and contingencies of the 'human condition', such as death, luck, conflict, human plurality, and alienation amongst others. As suggested earlier, green politics can, in part, be viewed as concerned with 'environmental crisis management', that is cultivating individual and collective modes of behaviour and character which help us realize our interests, and as suggested in the next chapter, encourage us to reflect on those interests as part of this process. Environmental virtue ethics differs greatly from deep ecology in that it does not seek to definitively answer or solve the existential riddle of human existence, or discover a permanent solution to social–environmental relations. At the same time possession and exercise of the virtues are also said to be required in order to live well. According to Aristotle, 'The good of man is an activity of soul in accordance with virtue' (1948: 1220a). In the classical formulation virtues are also integral aspects of the 'good life': that is the cultivating virtues, such as courage, prudence, charity and justice, are not simply instrumentally valuable but are also constitutive elements of the good life and human well-being. Thus, one cannot propose an account of ethics centred on virtue without also indicating what is the 'good' around which these virtues are orientated. However, as indicated in the conclusion to this chapter, and in Chapter 7, an account of the environmental virtues is compatible with a 'thinner' theory of the good than that usually associated with virtue ethics, such as that advanced by MacIntyre (1984) for example. A starting point for a 'modern' and green appropriation of the virtue ethics tradition is given by Geach who points out: 'A specific answer to the question of what men are for and what end they should aim themselves at is not required in order to show the need for the cardinal virtues' (1977: 45). As a praxis-orientated view of morality, a virtue ethics casts the green theory of the good in terms of being and 'doing' and not simply of 'having'. That is, the good of which green virtues partake is one in which human well-being is understood as constituted by action rather than possession or consumption. This theme is developed in Chapters 6 and 7.

The attractiveness of locating green moral claims within the idiom of virtue ethics comes partly as a reaction to the common misperception of green politics as requiring an excessive degree of self-denial and a puritanical asceticism (Allison, 1991: 170–8; Goodin, 1992: 18). While we may reject the claim that green political theory calls for the complete disavowal of materialistic lifestyles, it is true that green politics does require the collective reassessment of such lifestyles, and may require a degree of shared sacrifice. Some green policies, particularly in their initial stages, may be matters of necessity rather than intrinsically desirable. Initially therefore these policies have to do with things that are difficult for people, that is getting people to do things that they might not otherwise want to do, but can be demonstrated to be in their long-term interests. This is particularly the case in respect of ecological sustainability. This is where virtue comes in, since as Foot writes, 'virtues are about what is difficult for men [*sic*]' (1978: 8), i.e. correctives to human frailty, and weakness of will. Virtues, let it be clear, are *not* concerned with the elimination of human weakness. That is, while virtues do promote modes of being and ways of acting which will help individuals attain and discern their good, it is not part of the argument for virtues that they perfect the 'human condition'. To use ecological terminology, virtues may be thought of as character traits, modes of being which help to find the best 'adaptive fit' between the individual and her interests and the environment (both social and natural) she inhabits. The importance of virtues for the green position resides in the necessity of self-restraint, prudence and foresight so that long-term (i.e. sustainable) well-being is not sacrificed or undermined by desires to satisfy immediate self-interest. This, as argued throughout the book, is the essence of ecological stewardship. The latter may be thought of as the cultivation of those modes of character and acting in the world which encourage social–environmental relations that are symbiotic rather than parasitic. In other words, ecological virtues are related to social–environmental relations in which human self-interest and well-being are fulfilled by modes of interaction which minimize harm to the interests of the non-human world as much as possible, without sacrificing serious human interests.

We may think of environmental virtue as having to do with the refinement of moral discernment in regard to the place of nature as a constitutive aspect of the human good. The cultivation of environmental virtues can then be regarded as a matter of discerning the place nature has within some particular human good or interest. A more positive statement would be to say that those who destroy nature are motivated by an unnecessarily narrow view of the human good, and that 'what they count as important is too narrowly confined' (Hill, 1983: 219). In so doing the inherent plurality of the 'human good' is occluded. That is, forms of anthropocentrism which narrow the human good and human interests can be criticized as vices, or potential vices. At the same time, those who destroy nature also often have a mistaken appreciation of the 'seriousness' (Taylor, 1989) of the human interest or good in the service of which nature is destroyed. However, to reject anthropocentrism is not the solution, but is rather itself a vice of which we need to be aware. A virtue approach is thus anthropocentric in that its reference point is some human good or interest, but as argued in the next chapter, this ethical (as opposed to metaphysical) anthropocentrism is compatible with including considerations of non-human interests and welfare.

Conclusion

The general view of deep ecology presented here is that its concerns relate to seeking answers at the metaphysical level pertaining to the proper relationship between 'humanity' and 'nature'. However, one could argue that deep ecology starts out from a particular understanding of that relationship and then works backwards, as it were, to provide various arguments for that *a priori* position. So the strong critique of deep ecology is that it assumes a particular metaphysical understanding of the way in which humans and nature ought to interact. Crudely put, deep ecology seeks to persuade us that there ought to be a prima facie disposition in favour of non-interference with nature by humans. Dobson succinctly expresses this fundamental deep ecology position as seeking to 'turn the tables in favour of the environment, such that the onus of persuasion is on those who want to destroy, rather than those who want to preserve' (1990: 69). The developments within deep ecology from its initial concern with defending the intrinsic value of nature to its more recent ontological and psychological turn can be seen partly as different ways to provide justifications for this *a priori* position which privileges, *ceteris paribus*, the preservation of nature over its use by humans. However, deep ecology is also found wanting from the point of view of providing a basis for normative agreement for green politics within contemporary societies.

However, what can be taken from deep ecology and incorporated within green moral theory is its emphasis on the dispositional aspects of morality. Deep ecology viewed as a form of virtue ethics does have something to offer in establishing the normative basis of green politics. The idea of stewardship is as a mode of productive interaction with the non-human world which is sustainable and symbiotic and, while anthropocentric, is compatible with the policy implications of ecocentrism. It therefore represents a more defensible moral basis for green politics. This anthropocentric virtue account of morality, in which character is a mode of being and a way of interacting in the world (both social and natural), is further developed in the next chapter.

Notes

---

1 This touches on the religious theme raised earlier, in that for deep ecology it is through 'ignorance' rather than innate 'sinfulness' that we treat the non-human world as we do. These religious overtones of deep ecology cannot be overlooked, particularly when placed alongside its 'redemptive' character. This ignorance can be taken to include both scientific and metaphysical dimensions. For example, for some green ethicists, an increased knowledge of

the natural world, our continuity with and dependence upon it, is a necessary prerequisite for any 'environmental ethic' (Callicott, 1982; Rolston, 1982). However, for some deep greens, our ignorance needs to be remedied not through science but in returning to 'The Way', a form of knowing and acting which is the 'natural', perennial, harmonious path of human/non-human interaction (Goldsmith, 1991). There is a striking resonance with the biblical story of the Garden of Eden, where the ignorance of Adam and Eve (who had not yet been tempted by the devil to eat from the tree of knowledge) provided them with that prelapsarian paradise.

2 As Marcel Wissenburg has noted, 'The point is that the Self of the beholder is a collective self of all (parts of) Nature. Therefore it only has one point of view' (personal communication).

3 In deep ecology the social as a constitutive aspect of morality is often undertheorized, and may explain why it seems to favour a 'quietist' non-interference position *vis-à-vis* nature (Wissenburg, 1993) and displays more concern with individual relations to the world than collective ones.

3

Rethinking Green Ethics II: Naturalism and the Ethical Basis for Green Political Theory

CONTENTS

One of the most distinctive features of green politics is its insistence that human relations with the non-human world are a legitimate object of moral concern, a concern which greens claim has been largely neglected within contemporary moral theory. This neglect has typically been argued to stem from the human-centredness that characterizes Western moral thinking, which lends ethical legitimacy to human–nature practices which are exploitative and morally unjustified. The strongest expression of this rejection is the claim that anthropocentrism represents little more than arbitrary prejudice in favour of our species. Anthropocentrism is held similar to sexism and racism, i.e. a form of 'human chauvinism' (Routley and Routley, 1979) or 'speciesism' (Singer, 1990), an indefensible and arbitrary bias in our treatment of other species simply because they are not human.[1]

As discussed in the last chapter, many green theorists and activists see the transcendence of this human-centredness as the only way in which the 'moralization' of social–environmental relations can occur. In this chapter I further develop the defence of anthropocentrism, briefly indicated in the last chapter, as the strongest and most appropriate ethical foundation for green politics. It is not my concern to enter into detailed discussion of 'environmental ethics', although reference will be made to salient debates and issues. The principal task of this chapter is more limited: to outline a moral theory consistent with green politics. That is, I am concerned with the relationship between the moral and the political aspects of green theory, as opposed to environmental ethics considered independently of green politics.

The starting point of the defence of a revised form of anthropocentrism as the most suitable ethical basis of green politics turns on a threefold distinction between arguments which state that:

1. There is an ethical dimension to social–environmental interaction.
2. There are a variety of reasons that can be given for this.
3. Similar treatment of the environment can be premised on different reasons.

Outlining these distinctions and the different arguments they express opens up a space within green moral theory between 'anthropocentrism' and 'ecocentrism', a space for an alternative position in which this perceived opposition may be overcome. This position, which is a naturalistic form of 'weak' or 'reflexive' anthropocentrism, is sufficiently flexible to accommodate the normative thrust of the ecocentric concern with protecting the interests of the non-human world. Following Passmore (1980) and Holland (1984), I argue that calls for the necessity of a non-anthropocentric 'new ethic' upon which to ground green normative claims are misplaced. To a large extent familiar moral language and what one may call the conventional anthropocentric 'grammar of morality' are sufficient to the green task. In this chapter, anthropocentrism is defended from charges of 'speciesism' by rejecting the idea that it is an irrational or arbitrary bias towards conspecifics. Focusing on human nature, anthropocentrism is argued to be perfectly rational and non-arbitrary, while the coevolutionary development of the human species with other species cannot sustain a radical moral separation between humans and the rest of nature.

'Environmental ethics' is presented as distinct from 'human ethics', in that they are two interrelated spheres of the human moral realm, with different concerns and priorities. I argue that norms regulating human–nature relations *supplement* rather than *replace* those regulating human relations. At the same time, green ethical naturalism is cashed out as an *ethic for the use of the environment* as opposed to an *environmental ethic*. As indicated in the last chapter, seeking to establish the *a priori* protection of nature from human behaviour is wrongheaded and unnecessary. The 'moralization' of human–nature exchanges does not preclude purposive-instrumental relations. An 'ethic of use' which recognizes the moral considerability of parts of nature can act as a side-constraint not only on *how* humans can use nature, but also on *whether* we ought to use nature.

The Political Context of Green Normative Theory

That human–nature interaction is a matter of ethical concern I shall take as self-evidently true. Firstly, it makes sense to talk of an ethical dimension to human–nature affairs, i.e. it is *intelligible* to talk of such a dimension. Secondly, that the human treatment of nature is an ethical, and a political, matter is something for which there is both historical and contemporary evidence in terms of human–nature practices in the West.[2] From the point of view adopted here, namely elaborating an ethical theory sufficient to support a political one, the issue is not that people do not regard human–nature relations as an ethical matter, but rather the different *reasons* that are given, or can be given, to explain, justify and perhaps extend such concern. As I hope to demonstrate, there need not be one over-riding reason for the moral treatment of nature, such as its possessing intrinsic value or being enchanted or sacred.

Additionally, I also assume that there are a number of areas on which nearly all participants in the debate concur. For example, the greatest area of agreement between ecocentric and anthropocentric positions concerns the welfare of domestic animals. 'Animal liberationists', environmental ethicists and green humanists are all united in their condemnation of a range of basic and widespread social–environmental practices. These include factory farming, intensive rearing practices such as battery hen 'production', hunting for sport, the use of animals for human entertainment, and animal experimentation for the testing of cosmetics. The 'humane' treatment of animals and the reduction of 'unnecessary suffering' is something all sides can and do agree upon. Thus, as suggested in the previous chapter, there can be agreement in practice while there is diversity and disagreement at the level of justification.

At the level of ethical theory there is also widespread agreement between the two positions. For example, the distinction between moral agents and moral subjects is widely used within the literature on green moral theory. Regardless of the initial starting point, most theorists when suggesting guides to action maintain an ethically relevant distinction between moral agents (typically, but not exclusively, humans) and moral subjects (some humans, non-human animals and other parts of nature incapable of being held morally responsible for their actions). The usefulness of this distinction is developed later, but here I simply want to stress that despite their critique of anthropocentrism and the 'speciesist' charges of unjustified partiality to our own kind, even the most ecocentric of deep ecologists do not avoid concluding that the fact of being human does have evaluative importance and that 'species impartiality' is impossible to achieve in practice and constitutes an extremely morally dubious guide to action.[3] The main point I wish to draw from these areas of theoretical and practical agreement is that an effective green ethical perspective sufficient to sustain green political principles and policies ought to start from the position where most of the other environmental ethical theories end: namely, how ethical principles are translated into practice. It is the emphasis on environmental ethical *practice* (which is premised on a view of morality as centrally related to human action, as indicated in the last chapter) that both makes for the advantage of adopting an anthropocentric as opposed to an ecocentric position and signifies the importance of the political background against which this practice takes place.

The Necessity for Immanent Critique

Rather than develop an ethic in isolation from the political context in which it will be applied and its effects felt, as given by specific environmental policies or altered socio-economic practices for example, sensitivity to the political context is central to the ethical position developed in this chapter. This political dimension can be partly explained by the centrality of human interests to the ethical position being defended, which stresses the way in which human interests can and ought to frame human–nature relations. Here, however, the political dimension has to do with the more practical fact that green arguments seek to persuade citizens, governments and other political actors of the normative rightness of these arguments, and to ensure popular support for whatever environmental policies or practices are consistent with these moral principles. Thus attention to the 'political environment' has to do with both the securing of normative agreement for green claims and also the practical impact of green policies on the non-human world as the 'measure' to judge the success of green politics.

On the one hand there is the idea that green politics ought to be aimed at securing public support for green policies and practices, which is of course essential for the democratic legitimacy of green politics as argued in Chapter 7. On the other hand any agreement must begin from an awareness of the political nature of many of the existing moral injunctions concerning human–nature relations. Prime examples are the various laws prohibiting cruelty to animals in most countries. Building on what was suggested in the last chapter, I wish to make the argument that the heart of the green political project lies in the exposition of the *contradictions* within contemporary moral thought and culture, rather than proclaiming the *total crisis* of Western culture and the bankruptcy of its anthropocentric moral tradition. It is the supposed bankruptcy, coupled with arguments concerning the 'dangerousness', of anthropocentrism that leads to calls for a 'new' (i.e. external) non-anthropocentric environmental ethic. However, since anthropocentrism has not been demonstrated to be either bankrupt or a 'dangerous' orientation to the non-human world (Fox, 1990), the compulsion to search beyond anthropocentrism for an appropriate moral idiom loses much of its force. An immanent critique of anthropocentrism ought therefore to be the strategy adopted in order to achieve public support for the normative ends of green politics. For example, it would seem more likely that greens would secure normative agreement for their position by identifying discrepancies within the present normative underpinning of current human–nature interaction. If greens present their normative case in terms of the contradictory and/or the incomplete nature of the current dominant moral consensus on human–nature relations, the *immanence* of their position means that it would be both stronger (because non-anthropocentric accounts do not hold up under scrutiny) and expressed in a language readily understood by those to whom its message is addressed.

Green politics can be seen as an attempt to show the internal contradictions of current norms and as an attempt to persuade people of the rightness of an alternative perspective on society's received attitudes to human–nature affairs. A good example of this contradiction lies in what Midgley has termed a 'discrepancy in the sensibilities of our age' (1992: 29). An example of this discrepancy, according to her, is that

> The steady growth of callous exploitation is occurring at a time when our response both to individual animals and nature as a whole is becoming ever more active and sensitive. There is accordingly now a much greater gap between the way in which most of us will let a particular animal be treated if we can see it in front of us and the way in which we let masses of animals be treated out of our sight than has arisen in any previous state of culture. (1992: 29)[4]

In this respect the green moral argument is a call for not simply the *extension* but ultimately the *deepening* of moral concern and understanding. Green politics is thus concerned with the task of political negotiation with regard to the moral as well as the practical status of social–environmental interaction within the social order.[5] It is an attempt to resolve what it takes to be salient contradictions within the existing social order concerning the treatment of the non-human world.

My strong suspicion is that the various forms of environmental ethics, from animal rights to deep ecology, influence how individuals behave not because of but in spite of their central arguments. What I mean by this is that a non-anthropocentric ethic 'works' because it is regarded by individuals as congruent with and understandable within the 'speciesist' moral 'grammar' of contemporary society and moral reasoning. Environmentalism has political resonance because it is largely seen as complementary to conventional human-centred modes of moral reasoning. And as argued in the discussion of naturalism below, viewing environmental ethics through the prism of human interests and human-centredness is not merely 'conventional' (in that it is an accepted or 'traditional' orientation, but also capable of being changed) but is also partly a 'given' which any effective ethic that proposes to alter human behaviour must take into account. For example, the intelligibility and effectiveness of appeals to 'animal rights' lie largely in the *rhetorical* impact of rights talk as a readily recognizable way in which to convey demands for certain expected kinds of human behaviour in the treatment of animals. As Singer states, 'the question of whether animals have rights [is] less important than . . . how we think they ought to be treated' (1979: 197). It is not animal *rights* but human *duties* that underpin this attempt at moral persuasion.

Moral Extensionism and Moral Reasoning

'Moral extensionism' is the dominant practical strategy that non-anthropocentric environmental ethics offers as a guide to action. This is usually expressed in terms of the extension of the 'moral community' to include non-human entities. This understanding of extensionism is based on what I call a 'proprietarian' notion of morality in contrast to a 'relational' conception of morality which I defend in the rest of the chapter. A proprietarian view sees morality as largely turning on the possession or non-possession of morally relevant properties (e.g. sentience, rationality, consciousness, intrinsic value). A relational view, as its name suggests, sees relations as central to morality, and as having priority over proprietarian considerations.

From a political point of view, extensionism does not imply the application of intrahuman moral reasoning to the non-human world. Rather, it denotes that there are parts of the non-human world which are already embedded within existing social and moral relations, and that such relations are constitutive of the moral sphere. Extensionism is a process by which the range of human–nature relations which are subject to moral considerations can be extended outwards as a result of critical deliberation. The political force of extensionism comes from the fact that the recognition of a moral dimension to human–nature affairs is an accepted aspect of moral life. There is nothing new about the *process* of including certain parts of the non-human world within the remit of morality. It is a perfectly natural feature of all human societies – 'natural' in the sense of being uniformly present in all human cultures. What is new about environmental ethics is the particular entities, and categories of entities, to which extensionists propose to extend moral consideration, and the type of consideration extended. The issue is not whether to include human/non-human relations as being moral, but rather which relations and to what extent they are to be so included. It is important to point out that whereas proprietarian extensionism follows an 'outside-in' strategy (from the existing realm of 'non-moral' human relations to the environment to the 'moral', on the basis of an external, non-anthropocentric ethic, thus including the previously 'non-moral' within the extended 'moral' sphere), relational accounts move in the opposite direction. The latter seek to include non-humans by reflecting on existing social–environmental practices and their moral components, rather than basing the treatment of non-humans on the possession of capacities alone.

Unfortunately, current formulations of extensionism are narrowly focused on looking for morally relevant properties such as intrinsic value, or capacities such as consciousness or sentience, and then using these as the basis for deciding who or what is to be included in the moral community. This motif of the 'expanding circle' of moral concern is common currency within the literature (Singer, 1979; 1990; Regan, 1982; 1983; Rolston, 1988; Nash, 1989). On this gloss, environmental ethics sees itself as the modern heir of the struggle to expand the moral community. The historical evolution of morality is regarded as one in which those owed equal moral consideration progressively transcend tribal, national, male, white criteria, and reaches its fullest expression in the declaration of universal human rights. Both animal rights theorists and environmental ethicists claim they are simply carrying on this expansion to its logical conclusion (Nash, 1989: 16). Just as non-whites, the working class and women were 'liberated', accorded equal human rights and included as full members of the moral and political community (at least at the level of rhetoric if not in practice), so too will nature (or significant parts of it) be 'liberated' by the process of extending rights and the equal consideration of interests (Singer, 1990; Rodman, 1995).

The proprietarian understanding of morality holds that the most important issue is the specification of the 'moral community', and that it can be explained by reference to one or a set of characteristics. By adopting an *a priori* anti-anthropocentrism as the only secure basis for the moralization of social–environmental affairs, proprietarian extensionism moves in the wrong direction, from the non-human inwards rather than from the human outwards. A good example of this type of reasoning is Westra (1989) who argues for the subsumption of intra-human and social–environmental affairs under a single comprehensive ethic. For her:

> rather than start with a 'humans only' ethical perspective and then strive to broaden its basis to cover other entities . . . it seems preferable to start with an all-encompassing ethics, and agree that other, more specific (perhaps stricter) sets of ethical principles may well govern each group's interaction (including our own). (1989: 224)

This is a clear example of the 'outside-in' extensionism that characterizes most proprietarian views of morality, nicely expressed by Rolston's declaration that 'we must move from the natural to the moral' (1992:

135). Other examples include Singer's (1990) utilitarian argument concerning the equal consideration of human and animal interests, or Regan's (1983) deontological ascription of rights to animals on the basis that some of them are 'subjects-of-a-life' with interests that ought to be respected and protected. These extensionist positions turn on the claim that what unites us in a morally significant sense is not that we are 'natural beings' like other creatures, with our own particular nature. Rather, what we are argued to share with other species are particular capacities such as sentience, or being 'subjects-of-a-life'. The problem with this form of extensionism is that the complexity and richness of moral experience are reduced in the name of expanding the moral community.

According to current extensionist arguments, moral relations should be species-impartial and capacity-sensitive. It is important to point out that what I wish to criticize is not the latter but the former. It is not my argument that the possession of certain capacities is a matter of moral indifference in the determination of proper treatment. Rather it is that a view of morality in which having these capacities is of paramount significance offers a problematic conception of morality, one based on species impartiality. This has to do with the link between proprietarian accounts of morality and 'discovering' rather than 'creating' moral relationships in a similar manner to that described in the last chapter in criticizing deep ecology. Morally relevant properties are already in the world, waiting to be 'discovered' by human agents, whereas ethical relations are created and maintained by humans. Thus proprietarianism is a central plank in the attempt to move away from anthropocentric partiality.

The reason for the prevalence of proprietarian accounts of morality within non-anthropocentric ethics is that they can secure impartiality in a way in which a relational account cannot. Relational accounts are accused of 'speciesism', i.e. partiality to members of our own species. However, for capacities to come into play as side-constraints on the treatment of non-humans, there must be a prior judgement about how we relate to bearers of these morally relevant properties. An awareness of the properties humans and non-humans share can help to diminish the separateness between the two. However, decreasing the *differences* between humans and non-humans in this way cannot erode the fundamental moral *difference* between 'human' and 'non-human'. That is, a proprietarian account cannot fully capture the idea that we stand in a qualitatively different set of relations to others of our kind than we do to non-humans. Through the adoption of a relational view, the various differences and similarities between humans and non-humans can be accorded their proper weight in moral deliberation on the conduct of relations between them.

The distinction between moral agents and subjects or patients is at the heart of moral extensionist arguments. The extensionist proposition is to increase the class of moral subjects deserving of moral consideration. On the whole, extensionism does not question the fact that only human beings can be moral agents, and thus held responsible for their actions. One of the standard extensionist arguments is the 'defective humans' argument popular within the animal liberation literature. This argument holds that if we accord some 'defective' humans the status of moral subjects (i.e. morally considerable but incapable of full moral agency), and wish to avoid the charge of arbitrarily tying treatment to species membership, then it follows that it is the possession of capacities rather than species membership that widens the class of moral subjects and thus the moral community. That is, capacities not species membership determines treatment. No one capacity or set of capacities can be found which will include all humans and exclude all non-humans. In this way this standard argument purports to deflate the moral relevance of being a member of the human species, while still maintaining that only humans can be moral agents. It works by separating humans into agents and subjects and implicitly constructing morality as composed of three sets of relations: those between human agents and other agents, human agents and human subjects, and finally human agents and non-human subjects. Impartiality or equal consideration is the rule that governs the totality of relations between moral agents and subjects. To treat a human moral subject, who in terms of capacities is similar to a non-human, differently from the latter, is argued to betray a 'speciesist' attitude, akin to sexism or racism. A serious problem with this proprietarian view is its abstract, cold and calculative quality. In the end, moral practice, the phenomenology of the moral life, is not constituted by impartial relations between two sets of 'capacity holders', moral agents (some humans) and moral subjects (humans and some non-humans). Rather the richness and texture of morality require seeing it as rooted in determinate social relations between humans, which include social–environmental interaction. To be a human being means (constitutively) relating to other humans (both agents and subjects) in ways not shared with the non-human world (subjects or otherwise). The fundamental moral relation in which humans stand to each other is thus irrespective of the possession of capacities, as argued later.

Rationalism and Moral Reasoning

Ultimately, what moral extensionist arguments share is a common critique of that dominant strand in Western moral thinking which makes *reason* and the capacity for rational deliberation of cardinal significance in moral affairs. It is extensionists' attack on what one can call the 'rationalist' tradition within moral theory which leads them to search for alternative properties not exclusive to humans (such as sentience) by which moral considerability can be established and the 'moral community' extended. Yet while the excessive rationalism of contemporary ethics is criticized, the latter's proprietarian logic is maintained. Moral extensionism criticizes merely the choice of rationality as the capacity whose possession imparts moral considerability and constitutes the core of morality, not the idea that the specification of capacities is that core.

It is also the case that 'animal liberation' theories of the deontological or utilitarian sort actually end up endorsing the primacy of reason. This is because these theories work by rejecting particularity, but also non-rational aspects of moral reasoning such as sympathy, so as to ground species impartiality. Singer is at pains to point out that the inclusion of animals within the moral community is based on reason not emotion. As he notes in the preface to the 1975 edition of *Animal Liberation*:

> The ultimate justification for opposition to both these kind of experiments [Nazi experiments on concentration camp victims and contemporary ones on animals] . . . is not emotion. It is an appeal to basic moral principles which we all accept, and the application of these principles to the victims of both kinds of experiment is demanded by reason not emotion. (1990: iii)

As will become clear in the next and subsequent sections, the role sympathy plays in moral life in general, and in the operation of that part concerning our dealings with the non-human world in particular, is perhaps dismissed too quickly by Singer. In part, it is the integration of reason and emotion that a naturalistic ethic seeks. Singer's rejection of sympathy as a basis upon which to extend moral consideration to animals can be explained by his earlier confusion of 'sympathy' with 'sentimentality' as when he states 'This book makes no sentimental appeals for sympathy toward "cute" animals' (1990: iii). Sympathy is not just a disposition towards 'cute' or 'attractive' animals/humans, but is also a mode of being, being able to comprehend another's situation and to a certain extent see the world from their perspective.[6] Thus Singer, and other proprietarian non-anthropocentrists, end up with reason remaining the prime feature of morality, although one which is sensitive to capacities such as sentience or properties such as intrinsic value. However, the non-anthropocentric critique of the dominance of rationality together with the extensionist logic are positive aspects of the proprietarian moral position. In particular the critique of rationalism is something that can be built upon. In the next section I attempt to show how this critique can be worked up within a naturalistic ethical theory, which while being speciesist can accomplish much of what extensionism hopes to achieve.

The problem with the way in which the extensionist argument progresses is that it often does not make clear that it is *rationalism* and the use of reason as the exclusive demarcation between 'human' and 'non-human', and as marking the border between the 'moral' and the 'non-moral', as opposed to *anthropocentrism*, that is really at the heart of their critique. Rationalism and anthropocentrism are seen as synonymous rather than the former being seen as a particular conception of the latter.[7] In opposition to the rationalist tradition, those who write on the morality of human-animal affairs from what can be called a broadly 'Humean' perspective, which stresses the role of sentiments, emotions and instincts, offer a more convincing moral outlook for green politics.[8] Rather than decentre the human from an alternative account of morality, what is decentred and placed within its proper perspective is the role of reason within moral reasoning. What can be termed the contextualization of reason reconceptualizes morality such that the insuperable barrier between the 'rational' world of humanity and the 'irrational' or non-rational natural world is not so much transcended (since the naturalism developed below holds that being human is of central moral importance) as placed within its proper context. Whereas standard extensionist positions seek to demonstrate the continuity between human and non-human by the use of the various trans-human properties mentioned above, and standard rationalist accounts of morality are premised on the radical separation between human and non-human, the naturalistic position defended below attempts to reconcile the thrust of these positions by demonstrating that humans are a *differentiation within* rather than a radical *separation from* nature.

Naturalism and a Defence of 'Speciesism'

If we reject, as I believe we must, claims of a single definition of ethics, and instead focus on ethics as a practice embedded in human social life, we will be in a better position to account for the inclusion of social–environmental relations within the scope of morality. This will also be the case if we reject the argument that the moralization of social–environmental relations requires the discovery of a single ethical code. One of the problems with proprietarian extensionism is that it is insufficiently sensitive to the myriad ways in which humans interact with the non-human world, and to the varieties of human moral experience *vis-à-vis* that world. At the same time it cannot be supposed that there are no limits to moral extensionism. Constraints on the practical scope of extending moral consideration to the non-human world are expressed by Brennan who claims:

> Even if morality succeeds as a device for counteracting limited sympathies within the human community, it is unlikely to succeed as a device that will enable us to yield priority over human concerns and interests to the good of things 'natural, wild and free'. (1988: 30)

The argument here however does not seek to establish the priority of nature's interests over those of humans. Naturalism, in starting from general 'givens' of the human condition, works with an explicit awareness of the restrictions, boundaries and contingencies that mark all human experience, including moral experience. It was these features of the human condition, it will be recalled, that made a virtue-ethics approach an appropriate one to take. Here we can supplement this view by holding that attention to what it means to be human not only shows the continuity between the human and non-human worlds (thus overcoming the standard argument levelled against anthropocentrism concerning the separation between 'non-human' and 'human'), but also delimits the effective range of moralized human–nature relations. The naturalistic meta-ethical position adopted turns on the idea that being human counts for something and that speciesism or prima facie favouritism towards members of one's own species is neither an 'irrational bias' nor akin to sexism or racism as typically held by non-anthropocentric theorists (Routley and Routley, 1979). Rather, it is at the centre of any workable moral theory covering social–environmental interaction.

Naturalism implies attention to how the moral treatment of non-humans relates to what it means to be human, with the supposition that the particular type of 'natural being' we are impinges on how we ought to treat non-humans. 'Ought', after all, implies 'can'. Our nature circumscribes the range of choices, ethical and non-ethical, open to us, which is not to say it determines those choices as sociobiologists suggest. Here the working assumption is that before we can 'ecologize' ethics in the way most environmental ethicists and other moral extensionists intend, we must first 'naturalize' ethics. The supposition is that asserting the central importance of human nature in ethical reasoning is a necessary starting point from which to argue for the inclusion of other parts of nature. In the following subsection I defend the argument that being human *is* of central moral importance. The reason for this is that the use of the predicate 'human' to describe a being is not simply descriptive but carries prescriptive force. Cicero's remark that 'The mere fact that someone is a man makes it incumbent in another man not to regard him as alien' (in Clark, 1995: 318) captures the essence of this argument. In *describing* a being as 'human' one is also *prescribing*, albeit in general terms, the type of relationship and general treatment owed to such a being.

Human Nature, Nurture and Culture

In presenting arguments about how we ought to treat non-human nature, we must also address the matter of human nature. The two are, as I hope to show, inextricably linked. If green politics is the 'politics of nature', as a recent book indicated (Dobson and Lucardie, 1993), then *human nature* must have a central place in its analysis. This aspect of 'nature' has not received much attention within the literature, and may account for the incoherence between green moral and political claims. When greens emphasize limits and point out that we cannot transcend certain 'givens' of external nature, as in the infamous 'limits to growth' thesis (Meadows et al., 1972) they often forget that the same caution applies to the nature of humans themselves.

This concern with the role of human nature within moral affairs is motivated by Brennan's suggestion:

> in order to discover what sort of human life is valuable we must first consider what kind of a thing a human being is. Although there is, in my view, no complete answer to this question, we can . . . grasp one important aspect of human nature by reflecting on what are essentially ecological considerations. (1988: xii)

In adopting a naturalistic ethical position, I do not imply that the moral foundation of green politics rests on a strong and determinate account of 'human nature'. Rather, it highlights certain salient features of human beings which are more or less 'givens' that any ethical theory must take into account. This understanding of 'human nature' is *a posteriori* based on anthropological and other empirical evidence, from which a naturalistic understanding of the human species can be inferred. Thinkers who adopt such a naturalistic perspective include Midgley (1983a; 1995), Hampshire (1983; 1989), Benton (1993) and Clark (1979; 1982). One reason for adopting this approach is that it is appropriate to talk initially of the *human species* when discussing the questions of interspecies interaction. We are products of evolution as much as culture and convention, and as such our range of realistic choices is 'framed' not just by culturally defined norms, if only because these cultural norms are 'shaped' by our innate dispositions as social animals. *Ex hypothesi* 'we' are adapted to 'our' culture, which in turn is, at least temporarily, adapted to its environment.

The first level of our nature turns on certain universals of human biological/physiological constitution. These distinguish us as a species from the rest of nature. For example, that we are vulnerable to harm from certain non-humans and not from others is important. We do not and cannot stand in the same relationship to all parts of nature. For example, trees, unlike snakes or viruses, do not present the same kind of danger to us. At this basic level, our treatment of trees will consequently be different from how we treat snakes. Given our make-up, it is simply inconceivable that all of non-human nature can be treated and viewed in accordance with one master principle. Humans, after all, do not face an undifferentiated and homogeneous entity called 'nature' but face specific parts of it.

The second level of our nature relates to the centrality of culture in the determination of human nature. It is not 'humanity' but specific groups or societies that interact with determinate parts of nature. An example of this cultural dimension is that need fulfilment for humans goes beyond mere biological subsistence. For example, Benton points out, 'Proper human feeding-activity is symbolically, culturally mediated' (1993: 50). Culture is our species-specific mode of expressing our nature, or species being. As it is continuous with our nature as social beings, human culture does not represent a radical separation from nature, but can be viewed as our 'second nature' (Bookchin, 1986), emerging from, but situated within, the natural order. The importance of this has been expressed by Kohák who notes, 'Were culture a negation of nature, no integration of humans and nature would follow' (1984: 90). Human culture can thus be seen as a collective capacity of humans to adapt to the particular and contingent conditions of their collective existence, including, most importantly, the environments with which they interact and upon which they depend. Thus, culture is in part the particular mode by which humans adapt to their 'ecological niche', but not simply in the sense that cultures are somehow 'determined' by environments. Rather, it is in the additional sense that the mode of human adaptation to their 'ecological niche', and the expression of their 'species being', involve the *active transformation of their environment and the creation of their 'ecological niche'*. As Lewontin puts it:

> We cannot regard evolution as the 'solution' by species of some predetermined environmental 'problems' because it is the life activities of the species themselves that determine both the problems and the solutions simultaneously . . . Organisms within their individual lifetimes and in the course of their evolution as a species do not *adapt* to environments: they *construct* them. They are not simply *objects* of the laws of nature, altering themselves to the inevitable, but active *subjects*, transforming nature according to its laws. (in Harvey, 1993: 28)

Part of the reason for this is that the 'ecological niche' for the human species is extremely wide, as can be readily seen in the success of our species' colonization of the earth's surface. As the species nature did not specialize, we are unique in the range of ecological niches in which we can flourish, present ecological problems notwithstanding. At the same time, membership of a culture also expresses our distinction from others of our kind. It is not an undifferentiated 'nature' that we face at the first level, while at the second level it is not the 'human species' that is the proper subject of analysis. Rather what we should be concerned with are determinate social practices and individuals within particular cultural contexts facing more or less determinate parts of the environment.

Viewed from the perspective of the late twentieth century, these environments cannot be taken to be 'natural', i.e. independent of human influence, since often they have been transformed by past and current human behaviour. The significance of this cultural dimension, raised in the previous chapter, is that what green politics seeks is a cultural adaptation to altered ecological conditions. In other words, the human 'ecological niche' is both culturally *and* biologically/ecologically determined. These aspects of human being (biological, cultural and ecological) have been expressed by Benton as implying that 'Humans are necessarily *embodied* and also, doubly, ecologically and socially, *embedded*' (1993: 103). We must use nature to live and to flourish, but we do not react uniformly to it in either our instrumental-productive or our moral relations.

An important naturalistic indication of this is the presence of various morally significant categories in almost all human societies for describing the non-human world. Just as in the use of the predicate 'human', some of these descriptions of the non-human world carry prescriptive intent. For example, Diamond gives this explanation of designating particular animals as 'vermin':

> the notion of vermin makes sense against the background of the idea of animals in general as not mere things. Certain groups of animals are then signalled out as *not* to be treated fully as the rest are, where the idea might be that the rest are to be hunted only fairly and not meanly poisoned. (1978: 476, emphasis in original)

Similarly we can think of other animal categories such as 'pet' or 'food animal', which like 'vermin' both describe the particular relation in which that animal stands to us as well as prescribing the appropriate treatment. The point is the ubiquity of this normative 'background' that animals are not mere 'things'. Morally relevant categories in regard to the inanimate world are less common but conceptualizations such as 'private property', 'garden', 'national park', 'city' and in less enlightened times 'forest', 'uncharted lands', 'wasteland' and 'wilderness' are testament to the ubiquity and naturalness of human moral categorization of the external world.[9] This human concern with categorizing the natural world demonstrates that because we do not interact with an entity called 'nature' or the 'environment', there can be no single moral principle to govern this multifaceted and complex relation. In short, such moral categories and the process of categorization are partly *constitutive* of that relation.

We do not react to the world as disembodied centres of rationality, as rationalist accounts of moral experience suppose. Neither, on the other hand, do we react to the world in the same way as other animals do; this is one of the mistaken assumptions of strong sociobiological arguments. The complex of human reactions to the world cannot be reduced to either of these positions. We are unique in the range and variety of possibilities open to us. It is precisely this self-reflexively grounded capacity for choice that allows the possibility of moral concerns to be a factor permeating social–environmental exchanges. One way of looking at this from a naturalistic perspective is to note the difference between 'open' and 'closed' instincts (Midgley, 1995: 52–7). Closed instincts are fixed, while open ones are tendencies for certain general kinds of behaviour which are learnt by experience. It is the relative openness of our instincts with regard to how we flourish and interact that permits the possibility of non-instrumental considerations. Closed instincts refer to a limited but crucial set of 'givens'. Chief among these closed instincts is a preference for others of our kind, i.e. 'speciesism' is a feature of human nature. Midgley rightly rejects the common argument equating speciesism with racism and sexism in arguing, 'The natural preference for one's own species does exist. It is not, like race-prejudice, a product of culture. It is found in all human cultures' (1983a: 104). Thus speciesism is not a product of Western culture or Enlightenment thought: it is a ubiquitous (and complex) feature of human life. However, even as a closed instinct, the preference for one's own kind does not imply, in the human case, a callous disregard for other species.

Ironically, although environmental philosophers have traditionally urged us to recognize the continuities between ourselves and other animals, they have often underestimated the double-edged implications of this. Rolston, for example, in stating 'We do not derive ought from what is, but what ought must not be contrary to what is in nature' (1979: 25), seems unaware that this 'nature' includes human nature. Acknowledging the similarities between humans and the rest of the natural order does not guarantee the outcomes these theorists propose. Attempting to dissolve our uniqueness is the wrong way to go about analysing the question of the moral sense of human/non-human interaction. If we were to take seriously the environmental argument stressing our animal nature, then adopting a purely instrumental attitude to nature, which is the 'natural' attitude of all other species, unconstrained by anything more than prudence and efficiency qualifications, would be justified. It is our uniqueness in being both *a part of* and *apart from* nature that allows space for ethical concerns to influence social–environmental affairs (Barry, 1995c). Although it is important to stress our continuities with the natural order, as soon as we allow moral evaluations of our behaviour (and our behaviour alone) we set ourselves off from the rest of nature (Williams, 1992).

'She Does Not Know Humanity Who Only Humanity Knows'

As even a cursory knowledge of the evolution of human society and the histories of human societies indicates, relations between humans and parts of their environment, particularly domesticated animals, have never been entirely devoid of moral content. For example, in every human society we find social relations coexisting alongside productive relations between humans and non-human animals. A strong interpretation of this is Benton's statement that 'Humans and animals stand in social relationships to one another . . . It implies that non-human animals are in part *constitutive* of human societies' (1993: 68). We need not accept this strong version to see that human culture goes beyond relations between humans. Indeed one could go so far as to say that to interact with other species according to criteria other than those pertaining to material consumption is itself constitutive of human nature and culture. A person who, or a human culture which, treated other species as other species treat and interact with one another, i.e. with no moral dimensions whatsoever, would be strange to us.

This coevolutionary character of the human species is a powerful basis upon which to argue for the moralization of human dealings with the natural world, and to argue for the extension of moral reasoning to more of these dealings. It is not the case that our moral dealings with the non-human world demand a non-anthropocentric 'new ethic' (Passmore, 1980: 187). One reason for this is that the extension of sympathy and consideration to the non-human world, such that our use or non-use of that world is conditioned by ethical considerations, is part of what it means to be human. Midgley asserts that 'Our social life, our interests and our sympathy can and must extend outside our own species, but they do so with a difference' (1983a: 19). Sympathy, which requires contact with and/or awareness of others, is a powerful source of moral concern, a source which can operate over the whole range of human action, both with one's own kind and with other species. In other words, a human being who displayed no sympathy at all for any non-humans would be strange to us, i.e. atypical and highly unusual. As Callicott notes, 'a certain modicum of sympathy, concern and benevolence is humanly normal, very little or none at all is aberrant' (1992: 191). An ethical dimension to human dealings with nature is thus not something unusual or novel. Like other ineliminable aspects of the moral life, an ethical concern for regulating the use or non-use of (parts of) the natural world is a feature of all human cultures. That this can take a variety of forms within different cultural settings does not in any way count against its significance. As outlined above, the variety of human uses of the non-human world demonstrates the manner in which relations with that world are doubly constitutive for human beings. Firstly, these relations mark the (moral) difference between 'human' and 'non-human' (dealt with in more detail in the following subsection); and secondly, they are constitutive of the (cultural) difference between groups of humans.

An argument for ethical extensionism would therefore be less difficult and more in keeping with some basic and morally crucial facts of human moral life if it took into account and sought to build upon this constitutive role played by the social dimension to human uses of the natural world. An awareness that, to paraphrase, 'she does not know humanity who only humanity knows' would go some way to integrating the ecocentric demand to overcome the human/non-human, culture/nature dichotomy with a more defensible naturalistic alternative ethical position. However, for this to be achieved the emphasis on the continuity between human and non-human needs to be placed within its proper context. And as is argued in the next subsection, this context is the fundamental difference between 'human' and 'non-human'.

'Being Human Counts for Something'

One of the motivations for arguing that the fact of being 'human' carries with it evaluative import is to deal with the standard argument within proprietarian moral extensionism (and animal liberation in particular) which equates 'speciesism' with sexism and racism, and thus finds it equally morally unjustifiable. I think the obsession with finding properties upon which to ascribe moral considerability (often in the form of the 'rights' of non-human animals and other parts of nature) represents a narrow and incomplete view of morality. One way in which this view of morality is incomplete is its lack of attention to the role and function of feelings as motives and guides to action. As Gruen notes:

> If reason were the sole motivator of ethical behaviour, one might wonder why there are people who are familiar with the reasoning of Singer's work [on animal rights], for example, but who nonetheless continue to eat animals . . . reason is only one element in decision-making. Feelings of outrage, revulsion, sympathy or compassion are important to the development of concrete moral sensibilities. (1993: 151)

In many respects the standard argument comparing human 'defectives' to animals, for example by arguing that one should not experiment on animals unless one is prepared to see it carried out on a similarly placed human defective, is to have one thought too many. While rightly pointing out that the *differences* between humans and animals in terms of some morally relevant capacities and powers possessed by one and not by the other cannot sustain a clear and permanent boundary between the two, such accounts do not register the fundamental moral importance of the *difference* between 'human' and 'non-human'. When we use the prefix 'human', what this sometimes conveys is that certain basic types of treatment are called for. Using the term 'human' with prescriptive intent is thus non-arbitrary. There is typically a limited, but nevertheless significant, number of duties, usually negative, owed to another human being that are not owed to other beings. This is why denying another person as 'human' is usually a central part of radical discrimination against perceived 'aliens' and 'outsiders', and they may receive treatment according to codes fit for human/non-human relations. In many tribal societies, the appellation 'human' was often tied exclusively to the tribe, even when contact was made with other humans.

For example, the universal moral abhorrence of cannibalism, the fact that humans ought not (under normal circumstances) eat their dead, or amputated limbs, can only be explained and fully understood by the distinction between 'human' and 'non-human' (or perhaps 'not human enough' in the case of eating one's enemies). As Diamond points out, 'what underlines our attitude to dining on ourselves is the view that *a person is not something to eat*' (1978: 468, emphasis in original). And from this perspective, it is beside the point that the person in question is 'normal' or 'defective'; the so-called *mere* fact of being a zoological human is not so inconsiderable after all. Being human counts for something; it is to be a kind of being to which others of the same species stand in a foundational moral relationship. As Pickering-Francis and Norman point out, 'human beings may justifiably attach more weight to human interests than to animal interests, not in virtue of the supposed differentiating properties, *but because human beings have certain relations to other human beings they do not have with animals*' (1978: 127, emphasis added). The reason why humans do not usually eat or experiment on each other, 'human defectives' included, is not typically out of a concern for their (lack of) capacities or interests. Rather it is a basic moral fact of life that under normal circumstances relations between 'human beings', regardless of cultural membership, are or ought to be founded upon a set of moral considerations. These moral considerations are limited to duties such as not eating another human (alive or dead), non-interference, mutual aid, as well as other less precise concerns relating to some degree of sympathy and empathy, particularly with another's suffering, and perhaps most importantly a recognition that the other is a fellow human being with all that that entails. In this way the 'otherness' of the other, like 'merely' being human, is not really the case. The strangeness of another human is never total. We can move towards and understand humans in a way in which we can never understand and move towards the non-human world.[10] As Midgley notes, 'It is never true that in order to know how to treat a human being, you must first find out what race he belongs to' (1983a: 98). In the last analysis what separates the 'human' from the 'non-human', morally speaking, is a difference of kind and not just of degree. Other humans are never completely inaccessible or strange to us in the same way as the non-human world (including animals) can be.

At the same time, a naturalistic understanding of human beings carries with it the claim that it is natural for humans to care about their descendants in particular and future generations in general. One way of looking at this is to see care and obligations to the future as a 'natural duty' (Rawls, 1972). This feature of human nature is another 'given', non-chosen feature of human nature, but one that can and does work in favour of environmental arguments for sustainability and ecological stewardship. Care and concern for future generations, as Norton (1991) and de-Shalit (1995) demonstrate, do have positive environmental implications in terms of acting as a side-constraint on present courses of actions, and are a common feature of 'sustainable development'. This 'convergence hypothesis', in which long-term human interests (including a concern for future generations) are consistent with, and indeed require, environmental protection, is thus another way of expressing ecological stewardship, introduced in the last chapter.

Human Interests and an Ecological Ethics of Use

Despite appearances to the contrary, the common environmentalist critique of anthropocentrism is not that it denies *any* moral considerations being taken into account in human dealings with the non-human world. Even a crude anthropocentrism, i.e. one that sees no independent value in nature or values nature purely in economic/materialist terms, can be compatible with the imposition of *some* ethical limits on that instrumental exchange. That this treatment may be exclusively tied to human interests and concerns does not mean that the relationship is *ipso facto* non-ethical. The common misconception of non-anthropocentrism is to suppose that independent moral standing is a requirement for morally motivated treatment. Most commonly this independent moral standing is argued to reside in the intrinsic value of the non-human world. This underwrites a view of green politics for which 'The important point . . . is that it seeks to persuade us that the natural world has intrinsic value: that we should care for it not simply because this may be of benefit to us' (Dobson, 1990: 49). The crucial terms here are 'not simply' and 'benefit'. Care for nature may be independent of direct human benefit but it cannot be independent of human interests. Equally, that this care may involve considerations of human benefit does not drain that caring relationship of its moral character.

For example, by claiming part of nature as property people are obliged to treat it differently than if it were unowned. Individuals now stand in a different relation to that part of nature, because they now stand in a different relation to other humans. Thus, the treatment of nature viewed purely as a human resource can be guided, at least in part, by ethical considerations. In discussing the question of how we ought to treat the non-human world the focus should be on the evaluation of the *reasons* given for particular types of usage. Much of environmental ethics concerns itself with establishing that treatment be premised on the independent moral status of non-humans, rather than focusing on the primacy of the *relational* character of human/non-human affairs. One possible reason for this proprietarian view suggested earlier is the non-anthropocentric conviction that a human-centred environmental ethic, resting on human interests in and valuations of nature, cannot guarantee the *a priori* preservation of nature from human use that many deep ecologists and environmental ethicists see as the mark of any 'true' environmental ethic. Anthropocentric moral reasoning is held to be a precarious and insufficient ethical basis for the protection of nature. If, however, we reject the notion that an environmental ethic must be judged by whether or not it secures this *a priori* protection for the natural world, and instead see the job of any environmental ethic as regulating *actual* human uses of nature and identifying abuses, then anthropocentrism *per se* (as opposed to particular conceptions of it) need not stand accused of being part of the problem rather than part of the solution. It is the conviction of those who believe that non-anthropocentrism is necessary for an environmental ethic that leads to an emphasis on the non-anthropocentric powers, values or capacities, and which marks much of what I have called the proprietarian strand of environmental ethics. This non-anthropocentric ethic presents us with a picture of the world in which humans are disinterested valuers. The naturalistic anthropocentrism of an ethic of use sees humans as 'interested and partial valuers', and active transformers of that world. Because a relational view ultimately turns on human interests and concerns, it is viewed as capable only of an 'ethic for the use of the environment' as opposed to a genuine 'environmental ethic' (Regan, 1982), defined as an ethic which gives non-anthropocentric reasons for the protection of nature. What I wish to do in this section is to argue that an 'ethics of use' which regulates social–environmental interaction is a sensible ethical platform upon which actual, concrete human–nature conflicts and decisions can be resolved, and upon which green politics can base itself.

The central claim of an ethics of use is that an extension of human interests can achieve many of the practical outcomes desired by non-anthropocentrists but on a more secure basis, that of critically interrogating human interests in the world. This position starts from Norton's observation that 'A narrow view of human values . . . encourages environmentalists to look to non-human sources of value to justify their preservationist policies' (1987: 222). A broader view of human values and interests in the world may obviate the necessity for non-anthropocentric sources of moral concern. Part of this broadening process involves the examination of human interests. The reason for this is that this critical and self-reflexive process opens up the possibility of new *moral relations* between humans and nature within anthropocentric moral reasoning. The problem with most critiques of anthropocentrism and speciesism is that they are insufficiently sensitive to its environmental possibilities, especially in relation to the political defence and articulation of green policies and ideas. Although an ethics of use is by definition human-centred and related to human interests, it argues the green position on the basis of qualitative moral distinctions within human interests. The justification of a particular practice on the grounds that it fulfils a human interest is no longer considered as acceptable simply because it is a *human* interest. In other words, the fact that a particular use of nature fulfils a human interest cannot be taken as a decisive reason for either its initiation or its continuation in the same manner. It is this understanding of speciesism/anthropocentrism that deserves to be criticized as 'arrogant humanism', the idea that the mere reference to human interests is sufficient to justify morally any human use of nature. *An ecological ethic of use argues on the other hand that human interests are a necessary but not sufficient condition for the justification of human–nature relations.* For any human–nature relation to be fully morally justified the particular interests which that relation fulfils must be justified. This position begins from Midgley's conviction that 'however far down the queue animals may be placed, it is still possible in principle for their urgent needs to take precedence over people's trivial ones' (1983a: 17). In short, not all human interests, simply by virtue of being human, are equally acceptable or justifiable. That we must consume parts of nature to flourish, and use it in other ways to mark human life, does not mean that all uses are equally justified. Some are more morally defensible than others. The aim of green politics then becomes centred on determining defensible or permissible human uses of the non-human environment, and distinguishing these from impermissible, trivial and unjustified abuses.

Human Preferences and Interests: the Good, the Trivial and the Wanton

Adapting to my own purposes a distinction Norton (1984) draws between 'strong anthropocentrism' and 'weak anthropocentrism', I now outline the way in which an ecological ethic of use turns upon the distinction between human preferences and interests as well as the extension of human interests. According to Norton, 'A value theory is *strongly anthropocentric* if all value countenanced by it is explained by reference to satisfactions of *felt preferences* of human individuals' (1984: 134, second emphasis added). For my purposes the distinguishing feature of strong anthropocentrism is the claim that human–nature relations can be justified by reference to felt or 'given' *preferences alone*. From the strong anthropocentric position it does not make sense to talk about moral judgement derived independently from human preference fulfilment. It is the reductive character of strong anthropocentrism in conceiving of human–nature relations in purely instrumental terms (typically economic) that 'crowds out' both the need for their justification and the requirement that moral considerations ought to act as side-constraints on how relations are managed. Strong anthropocentrism, in holding that the moral justification of human uses of the natural world need not go beyond reference to preference fulfilment, reduces the various human interests that extend (or could extend) over our relations with nature to the preferences humans happen to have. According to Norton, if preferences are insulated from critical appraisal, human interests in the non-human world are in danger of becoming narrower than they might otherwise be.

At root, this green critique shares with critical theory the view that economic reasoning, if unchecked, can lead to the 'demoralization' of human–nature exchanges. Green politics is thus not against economic reasoning but rather seeks to place it within its proper context, as one amongst other modes of human interaction and valuation. From a green point of view, the aim must be to assess preferences by reference to the 'seriousness' or 'worthiness' of the human interest to which it is related. An ethics of use is thus a form of 'weak anthropocentrism', which acknowledges a plurality of human interests in the natural world and a variety of possible (but limited) morally appropriate relations to that world. It explicitly sets itself against any one 'true' relation to the world, and acknowledges the 'tragic' dimension of human being in the world in the sense of fully recognizing that, as in other areas of human life, human–nature affairs are often characterized as a clash of competing and mutually exclusive 'goods'.

As far as human–nature affairs go, the capacities of non-humans do count for something, for they partly identify what a being is, and are indispensable in guiding us in how to treat it; but as argued above, in the human case it is the *relations* between individuals that mark the difference between human and non-human. Therefore, while a proprietarian account is important for knowing how to treat a non-human, it is only one among a number of considerations in the human case. Proprietarian considerations, discussed in the next subsection, are largely parasitic on the prior justification of the particular practice in question. Even where a practice is deemed to be morally justified, attention to such considerations may require an alteration to the way in which it is carried out. Relational considerations, which hinge on assessing interests, address the issue of *whether* specific human uses should continue, while proprietarian ones generally bring up issues around *how* that relation ought to be conducted. An ethics of use is concerned with establishing the contested boundary between legitimate use and abuse, as well as the often more complex issue of when use cannot be morally justified, i.e. the line between use and non-use.[11] These boundaries can never be fixed in the manner that an *a priori* commitment to the preservation of nature would require, but are ineliminably contingent. This relative indeterminacy of this ethics of use argument will not satisfy those who seek cast iron protection of the non-human world. However, there is simply no remedy for this, and the most we can do is to acknowledge it and make it explicit. This is the responsibility of green politics, in that the implementation of green policies is partly a matter of 'monitoring' this indeterminacy and identifying from case to case when and why human use of the environment becomes unjustified abuse.

In environmental ethical discussions phrases like 'for no good reason' or adjectives such as 'wanton' are of critical importance because they indicate that morally wrong acts against non-humans are wrong because they cannot be justified by reference to good reasons. For example, common sense moral reasoning holds that one ought not to cause unnecessary suffering or avoidable harm. The keywords here are not only 'suffering' and 'harm', which pick out the relevant non-humans that are morally considerable, but also the intentional or voluntary dimension of the description as expressed by terms such as 'unnecessary' and 'avoidable'. Here the question is with the motive of the action. Typically, it is the 'wantonness' or 'triviality' of an act that merits moral censure. The capacity to suffer or be harmed is important in specifying the injured party, but it is the character of the act and/or the actor that is the focus of moral judgement. For example, slaughtering an animal may seem wrong but if, given the circumstances, it can be shown that it possesses ritualistic significance, is absolutely necessary for a group's sense of identity, and does not cause the animal an excess of pain and suffering, then it *may* be justified. But the same act carried out simply for 'fun' would not carry the same weight.[12]

One of the distinguishing features of the ethics of use is that it stresses the rather obvious idea that the participants are not an undifferentiated 'humanity' facing an equally undifferentiated 'nature'. In respect to the latter, there are morally relevant distinctions between domesticated animals, other animals, other living beings, plants, and inanimate nature. There is no one overarching principle which will cover all human relations to these various parts of the non-human world. Nor can it be expected that we should treat all natural entities within a particular category in a similar manner.

An ethics of use is particularly suited to issues relating to human dealings with domesticated animals, where 'abuse' can take on the more visible and readily identified form of suffering. It is within this category of human relations with animals that sympathy is strongest, and where we may expect greatest political agreement for altering these relations. This is because we can sympathize with other social animals, their pain and suffering, in a way which is not possible with some other animals and the vegetable or inanimate worlds. That is, their lives are intelligible to us in a way the modes of existence of these other parts of nature are not. Despite its rhetorical attractiveness, for example, it is difficult, to say the least, to 'think like a mountain'. Another important factor in the extension of sympathy to domesticated animals is the (relative) publicness or visibility of our treatment of them. Although the majority of people do not have day-to-day contact with domesticated animals (except of course as ready-wrapped meat, milk or leather products), there is some, although incomplete, awareness concerning the relationship. And as the last two decades of animal welfare activism demonstrate, the more people know about 'food' animals and how they are treated, the greater the sympathy with them and the demand for more 'humane' treatment of them, often independent of moral arguments.

A good example of the application of an ethics of use is the practice of factory farming. The latter reduces animals to the status of pure resources and denies them any opportunity to express their social, i.e. 'natural' character. It can be, and increasingly is, judged as morally wrong. As Benton argues, 'This form of "humanism" [factory farming] conceptualizes the needs of animals as instinctual and fixed in a way which simply leaves no room for morally significant differences to emerge between existence and thriving or living well' (1993: 59). There are alternatives available that would permit meat eating to continue but under more 'humane' production conditions. As indicated in the last chapter, animal rearing, viewed as a social practice which contributes to stewardship, represents a form of human productive relation with non-humans which is less 'abusive' than factory farming. From an ethics of use position, the suffering inflicted upon such animals cannot be justified simply by appealing to the higher economic costs associated with alternative husbandry methods. The human interest in food, and indeed, meat as food, can be satisfied in other, more humane ways. In other words, the suffering of these creatures is unnecessary. Factory farming as a particular form of productive relation between humans and animals is abusive, and their suffering unjustified. In the case of factory farming the appeal to human interests in profit-making or having cheaper meat is insufficient. For Benson the case against factory farming is based on 'the wholesale torture of animals that goes on in the name of nothing that could be regarded as a serious human purpose' (1978: 530). Thus in some circumstances human preferences (and perhaps in more limited set of situations, interests) ought to be accorded less weight, such that the welfare of non-humans may be considered to be of more pressing moral importance.

Morality and moral reasoning have to do with something serious (Foot, 1978; Midgley, 1983a; 1983b). It is this quality of 'seriousness' in relation to human interests that is crucial in the reordering of human priorities that moral extensionism entails. Some reasons for using nature in particular ways simply do not carry as much weight as others. Uses of nature which involve its material consumption and are not for serious reasons deserve to be morally condemned. Thus in regard to the factory farming example above, what makes it morally wrong is not that it is justified in terms of *human* reasons, as the deep ecologist and non-anthropocentrist hold. Rather it is wrong because the particular set of circumstances surrounding its operation render its justification on *economic* considerations alone as less deserving to count as a *serious* human interest.[13] There are alternative ways in which such preferences may be fulfilled, and although it is possible that some preferences will not be satisfied, it is not the case that abolishing this particular practice would result in great harm to central human interests such as political liberty, social justice, need fulfilment above subsistence, or the minimization of human suffering.

Thus an ethics of use indicates another important division pertaining to an additional set of limiting conditions on the operation of any green framework or process regulating our dealings with nature. This is the division between possible effects on human welfare of policies and altered practices issuing from this green process on the one hand, and considerations relating to human liberty on the other. This is discussed in Chapter 6. Without going too deeply into the issue, what I wish to signal here is that environmental policies which negatively affect human welfare are more defensible than those which compromise fundamental human liberties. This distinction is useful because it not only reinforces that there are different spheres of ethical inquiry (human-only and human–nature) but suggests some (admittedly inexact) 'rules of engagement' in the case of conflict between the two spheres. Thus social–environmental relations concern the trade-off between human and non-human welfare alone. What this means is that environmental policies which threaten central human liberties, such as the right to elect and influence government, or formative elements of human identity, for example need fulfilment beyond subsistence, are deemed illegitimate and prohibited in any but the most pressing of circumstances. In this manner green politics accepts that trade-offs in any conflict between the two ethical spheres are limited to possible sacrifices in human economic welfare and what Norton calls 'consumptive values' (1984: 135). This priority rule is widely endorsed even within non-anthropocentric positions when it comes down to questions of implementation and actual policy proposals. Unfortunately, as this rule is usually acknowledged at the conclusion rather than the start of most environmental ethical deliberation, its importance goes largely unnoticed. In the terms outlined above, in the case of these sets of foundational and non-negotiable human interests the question is *how* nature is used rather than *whether* it is used. On the other hand an ethics of use attempts to delineate the ethical threshold beyond which the human use of nature becomes abuse. In other words, the fulfilment of human liberty and welfare interests is compatible with ecologically sensitive use as well as principled non-use. The point of the ethics of use is to find symbiotic rather than parasitic social–environmental relations.

Ethics of Use and Ecological Virtue

Having derived a more focused ethic of use from a naturalistic perspective, we finally come to look at how this ethic of use relates to green politics. In this section it is argued that the idea of 'ecological virtue' captures the general thrust of the part such an ethic can play within green political theory and practice. Unlike environmental ethicists like Westra (1989) who argue for a 'new ethic' sufficiently broad-ranging to include intrahuman as well as human–nature relations, an ethic of use for the environment maintains that there is a fundamental distinction between these two sets of relations and spheres of moral life. Her position (which typifies much of North American environmental ethics) is that 'the main thrust of any ecological ethic is that it refuses to accept different sets of values, one for an isolated community of human beings and another ranging only in the wilds . . . we *can* have *one* extremely broad ethic' (1989: 224, emphasis in original). An ethic which ranges in the wilds seems to indicate that relations between non-humans are somehow 'ethical' in themselves independent of humans (which seems to suggest that the moral 'policing' of 'natural' ecological relationships is part of green politics). This section seeks to develop further the argument that the search for one ecological ethic is both theoretically impossible to sustain and unnecessary for green political prescriptions.

As set out below, separating out moral concern into intrahuman and human–nature spheres is a necessary precondition for establishing the normative claims of green politics. This separation is not a simplistic and regressive rephrasing of the human/nature duality, but represents part of a wider process transcending that duality. Green politics, in basing itself upon a view of morality in which virtues are central, seeks to create what may be called 'symbiotic' rather than 'parasitic' social–environmental relations: that is, the cultivation of ecologically virtuous modes of interaction with the environment. The border between *symbiotic* and *parasitic* relations denotes the ethical border between use and abuse.[14] Certain dispositions of character, and their social requirements, are held to be constitutive aspects of human–nature relations. Some of these virtues are peculiar to this domain of moral life, but some, perhaps the most important, such as sympathy or humanity, range over all aspects of morality.

The point about the virtues is that they are both partly *constitutive* of human well-being, while also being instrumentally valuable to human well-being. The claim that the practice of ecological virtues is constitutive of well-being helps to offset the notion that green politics necessitates sacrificing human welfare, a common view premised on the idea that the relationship between humans and nature is necessarily a zero-sum game. This leads to parasitic forms of social–environmental relations. Green politics, in the search for symbiotic relations, can take inspiration from the deep ecology notion that while human flourishing is compatible with decreasing human impact on the world, the flourishing of the natural world requires such a decrease.

A related issue concerns the common misperception regarding an apparent conflict of obligations between those in the human–nature ethical sphere and those in the human sphere. The usual way in which this is presented is that a moral concern for non-humans, particularly animals, means that there is less concern, compassion or sympathy 'left over' for humans. However, a virtue ethics perspective can easily demonstrate the wrongheadedness of this assumption. The practice of ecological virtue does not decrease compassion and concern for humans but rather the opposite. As Midgley points out, 'concern, like other feelings . . . is something that grows and develops by being deployed, like our muscles, not a sort of small oil well that will run out shortly if it is used at all' (1992: 35).

Citizens, Consumers and Ecological Virtue

A political interpretation of the ethics of use turns partly on a shift within the 'human' part of the human–nature relation. Green politics, if it is to have any purchase on the real world, any input into aiding the complex process of how real environmental problems are to be resolved, must take as its subject matter the actual, tangible relations and practices between particular humans and particular environments. Naturalism indicates the general background and limiting conditions of social–environment relations, while an ethics of use indicates that a 'citizen–environment' perspective is the most appropriate standpoint from which to judge *politically* the normative standing of the non-human world.

This perspective introduces what can be called a 'green conception of citizenship', dealt with in more detail in Chapter 7 where it is presented as a key issue in the relationship between green politics, democracy and the state. For the moment, all I wish to note here is that from the ethical standpoint the practice of the 'ecological virtues' is constitutive of this green conception of citizenship. The importance of this conception is that the practice of green citizenship can be regarded as the process by which individual preferences may be transformed not just as a result of reflection, justification and debate, but also because the virtues educate and refine preferences.

In part this process has to do with the idea that individuals as *consumers* have interests which are different to those they have (or potentially have) as *citizens*, and that on the whole, 'ecological interests' are not well served by the former.[15] Ecological interests here encompass both the interests of the non-human world, and human interests in that world. As noted above, economic reasoning *if left to itself*, in simply taking preferences as given and beyond evaluation, is more likely to result in practices which will unnecessarily harm non-human interests. At the same time if economic preferences are the sole or primary way in which human interest in the environment is expressed, non-economic ecological interests are less likely to be articulated. In other words, *qua* consumers in the market, individuals have a narrower set of ecological interests than would be the case *qua* citizens since, as citizens, their preferences are only provisionally 'given', are not immune from critical debate and discussion and are capable of being transformed. Simply put, as consumers people have a more limited range of interests in the environment than as citizens, principally turning on the economic benefits of ecological systems.[16]

The terms 'consumers' and 'citizens' are also shorthand for distinguishing market and political approaches to environmental problems. Aggregative, market-based approaches to social–environmental problems often fail to register the 'public goods' dimension of the environment. That is, one cannot ascertain the value of the environment as public good simply by aggregating private valuations, since these valuations will be dependent upon other people's valuations and the mechanism through which the public good is delivered. Consumer approaches to environmental issues also fail to register correctly the ethical dimension of social–environmental issues. The consumer mode of acting, the consumer character as it were, is inappropriate to articulate, never mind integrate, the range of human interests in the non-human world. The appropriate idiom, according to Jacobs, is one in which 'what is done to it [the environment] can be discussed in terms, not simply of costs and benefits (whether private or public), but of right and wrong' (1996: 1). Part of this relates to the fact that within modern society, new forms of social–environmental relations as a result of technological improvement have outstripped our moral capacity to deal with them. In this sense green politics seeks to point out that this power needs to be balanced and regulated by a collective sense of moral responsibility. That consumer-centred or market-based approaches to social–environmental problems restrict the range of operative human interests in general, and in particular often reduce interests to preferences, is a fairly standard criticism (Dryzek, 1987; 1996; O'Neill, 1993; Jacobs, 1994; 1996; Norton, 1994).

The articulation and creation of expanded interests is tied up with the collective political deliberation on the 'ecological common good'. There is no reason why this common good, a central concern of 'classical' views of citizenship, should be narrowly restricted to the human good. It is the willingness to search for ways in which human interests can be fulfilled with minimum harm to the non-human world that is one of the traits of green citizenship.

The point, however, is not to prioritize the citizen over the consumer as a mode of being, or indeed a site of ecological virtue. Rather the aim should be to integrate and harmonize them as far as possible. The importance of stressing citizenship is in part to acknowledge that any *moralization* of human–nature relations is directly related to their *politicization*. It may well be that there will never be consensus on the metaphysical basis of our moral dealings with nature. From a political point of view, pitching one's argument at this abstract metaphysical level is neither necessary nor helpful, since it sets green politics an unrealistic task. As a political theory seeking to persuade others of the normative rightness of adopting symbiotic (as well as sustainable) forms of social–environmental relations, green politics does not need to insist on convincing people of the intrinsic value or rights of nature. Rather there are a variety of ways in which we can explore the moral dimensions of our dealings with nature, and the search for one, overarching 'environmental ethic' is unhelpful to green politics. The issue of the moral status of our collective relations to the environment cannot, I believe, be resolved at a philosophical level, but can only be resolved politically, and even then only provisionally.

Ecological Virtues and Moral Character

The majority of the ecological virtues concern avoiding ecological vices. Such vices include familiar ones like inflicting unnecessary harm or suffering to animals, particularly domesticated animals in our care, and wanton and irreversible destruction of other parts of nature such as ecosystems. Ecological virtues aim at encouraging (not guaranteeing) that the human use of nature stays within morally justifiable limits and does not become morally unjustifiable abuse. The cultivation of the ecological virtues, the creation of 'ecological character' and dispositions, help create and maintain a proper balance within social–environmental relations. The emphasis on character stresses the importance of cultivating dispositions and modes of action which will discourage acting from 'wantonness' or ignorance. In avoiding the latter vice, scientific knowledge is of course propaedeutic.

This discriminating aspect of the ecological virtues picks up that stream of environmental ethical theory which emphasizes the relation between ecological science and environmental ethics (see Rolston, 1982; 1988; Norton, 1984; 1987; 1991; Callicott, 1992). Although the attempt to 'read off' how we should treat nature from ecological observations is fundamentally flawed, the ways in which ecological science and environmental ethics have been connected leave much to be desired. Ecological virtue suggests a possible way of resolving this issue by making the virtues appropriate to human–nature exchanges a matter of moral character. Thus, following the classical tradition, we can say that the ecological virtues are a combination of 'intellectual' and moral virtues which together combine to foster character. Both contribute to the cultivation and development of moral character. Green moral theory then becomes a theory of 'ecological character', a particular mode of behaviour, a way of knowing, feeling and acting. For example, we would expect an ecologically virtuous individual to be sympathetic, as opposed to sentimental, towards the non-human world, and to act from knowledge and experience of the world rather than from ignorance.

This division of the ecological virtues pertaining to moral character along traditional lines reinforces the central moral importance of adopting a discriminating attitude to the natural world. Ecological character can be viewed as an integrative and integrating mode of acting and thinking in which other modes (such as citizen, consumer, parent, producer) are combined. Depending on which part of that world is under consideration, either, or in some cases both, sets of virtues will be called for. For example, it is perhaps with those non-human animals closest to us, both in terms of familiarity and/or proximity and especially in their social nature, that the virtues of character will be strongest, as indicated earlier. In these cases, 'reason' or moral reasoning does not need to inform us that such creatures and their suffering are proper objects of our sympathy. It is simply the case that with such animals as domesticated cats, dogs, cattle and horses, and non-domesticated animals such as apes, dolphins, whales and elephants, their social nature makes their lives intelligible, however dimly, to us and their sufferings more recognizable, such that it is usually quite easy to sympathize with them. Such extended empathy becomes progressively more difficult to sustain as we move to other living creatures such as slugs, insects or rats, to plants and finally to the inanimate world of rivers, seas, mountains, rocks, ecosystems and bioregions. Although we should not discount the possibility of empathy with the inanimate world (for example, there are those who attempt to follow Leopold's advice and 'think like a mountain': Seed et al., 1988), it is more likely that as we move through the various categories of nature the virtues that inform appropriate treatment will emphasize those pertaining to the intellect.[17] Thus, a lack of sympathy with the inanimate world is not an insuperable obstacle in terms of cultivating a proper regard towards it. It may not be felt as intensely as moral sentiments in respect to fellow social mammals, but ecological science as a component of intellectual ecological virtue can extend our moral interests (if not our sympathy) beyond the animal boundary. Science can thus inform us of new and proper objects of moral sympathy and concern and thus of new moral relationships with wider and more distant parts of the natural world. For example, ecological science can inform us that what on first sight looks like a 'barren' wasteland is in fact a rich and thriving ecosystem (O'Neill, 1993). Indeed, to see it as 'barren' is to look at it purely from a particularly narrow perspective. Scientific knowledge can thus expand and refine our perspective and, perhaps, our interests in the world. That is, scientific knowledge can be a corrective (a virtue) to the vice of seeing nature from an unduly narrow and uninformed viewpoint, while also acting as a corrective to potentially harmful sentimental, ignorant or naive views of the non-human world.

One example of ecological vice can be found in the increasing gap between human transformative relations with the environment and the consumption of the goods and services created by those relations. It seems that the greater the gap between production and consumption the greater is the likelihood of parasitic rather than symbiotic productive relations. Such a view is consistent with the 'corruption' of social–environmental practices, such as farming and animal husbandry, owing to the imposition of external institutional norms as a result of the transformation of these activities along industrial lines. Once this occurs the internal norms and excellences which regulated the treatment of their non-human subjects are no longer operable. Equally, these parasitic, i.e. abusive, forms of productive relations arise and continue partly because consumers are largely unaware of them, and their lack of direct experience or awareness of animals within these productive relations limits their sympathies and thus their moral concern. At the same time unsustainable forms of productive relations come about partly because of the physical distance between the sites of production and consumption. An important aspect therefore of ecological moral character refers to attempting as far as possible to bridge this gap, both materially (in terms of 'prosuming' and self-reliance as discussed in Chapter 6) and morally (by extending moral sympathy and humane treatment to non-humans involved in these productive relations). However, this is not to condemn consumption and the consumer mode of human experience, thought and action as necessarily ecologically (and morally) wrong. The point after all is to search for the integration of those modes which affect the non-human world. Consumption is viewed as not merely an individual activity but a modern mode of being which has a definite social and cultural dimension, and the aim of green politics is to rethink or reconsider this mode without necessarily rejecting it.

Conclusion

The position outlined in this chapter presents not only an alternative green ethical theory to those advanced by non-anthropocentrists, but also an alternative understanding of morality and the moral life to that found in non-anthropocentrism. The main features of this alternative view are that morality has to do with action and experience, relates to character, and is divided into two broad spheres, one pertaining to humans only and the other to human/non-human relations. This *relational* view of morality is indispensable to the development of a feasible green moral theory concerning human–nature affairs. Unlike a proprietarian view it begins from where we are now, and through an immanent critique and engagement with humanism seeks to develop it in a more reflexive, less arrogant form. Thus the ethical dimension of green politics can be viewed as a form of 'moral extensionism', characterized as (a) weakly or reflexively anthropocentric, (b) naturalistic, (c) premised on an understanding of morality which sees it in relational terms, (d) turning on the importance of human and non-human interests, (e) not seeking 'one new ethic' to cover all spheres of human action, and (f) stressing the cultivation of moral character as the best way in which to integrate this reflexive anthropocentrism.

The relational view suggests that morality is to be understood as a practice rather than a catalogue of principles. In a sense there can never be a complete rendering of the 'ethical'. Following Aristotle, we can say it cannot be completely systematized (Clark, 1982). Principles cannot therefore 'map out' the ethical realm in full: it is something to be created in and through experience, rather than discovered. However, this is not to say that morality is created *ab nihilo*, or that it can be about anything. From the naturalistic point of view our status as a particular species standing in certain relations to each other, not shared with non-humans, is constitutive of what morality is about. Of particular significance is the naturalistic claim that humans do care about their descendants and also that care for the non-human world is another natural feature of human societies. This future-orientated concern can be seen as a central part of responsible stewardship. The strength of the naturalistic perspective may be said to lie not so much in the ability to provide definite answers to moral predicaments between the human and the non-human worlds. Rather it provides the appropriate 'moral grammar' in which to articulate those questions.

The ethics of use stresses what one may call the 'ecological management' concern of green politics. The normative commitment of green politics in terms of human–nature relations can be understood as placing individual human interaction with, and transformation of, the environment within a moral and political context. The apparent overtones of hubris and arrogance associated with the idea of 'ecological management' are just that, apparent rather than real or necessary. An ethic of use is consistent with principled non-interventionist solutions to human–nature conflicts. Thus, it can realize the 'hands-off' goal of deep ecology, as discussed in the last chapter. Set within this moral context, green politics is concerned with the provision of informed, ethically appropriate, and flexible social 'coping strategies' in relation to the ever shifting horizons of the relationship between the human and the non-human worlds. Human use of the non-human world, to mark as well as maintain human life, does not necessarily lead to the degradation of that world, and, despite local conflicts of interests, there can be a large degree of harmony between human and non-human interests.[18] The point of ecological virtue and the cultivation of an ecological moral character is that this harmony of interests *ought to be*. However, as suggested earlier in this chapter, and to be developed in Chapter 5, any harmony between human and non-human interests must be congruent with the fact that the human 'ecological niche' is the outcome of a metabolic, transformative relation between culture and nature. Thus any harmony between the human and the non-human worlds is 'created' not 'discovered', a product of culture not a natural process of nature. The character traits that most fit with this view are those of stewardship, a mode of transformative interaction with the non-human world in which this harmony is realized. However, as a mode of interaction it is not based on ignorance, arrogance or 'quietism', and it orientates itself towards establishing a 'respect' rather than a 'reverence' for nature. It seeks to avoid these ecological vices through the cultivation of ecological virtues, adaptive dispositions and habits which recognize the legitimacy of human transformation of the non-human world, but encourages a self-reflexive, discriminating attitude as regards that transformation. The fulfilling of some human interest is a necessary but not a sufficient condition to justify any use or transformation of the non-human world.

Green theory is premised on the normative rightness of transforming morally wrong or unworthy 'parasitic' social–environmental practices into 'symbiotic' ones. The 'ethic of use' complements weak anthropocentrism by acting as a first line of defence for non-human welfare in the clash between human and non-human interests. Or rather, in this clash human interests are reorganized in such a way as to create a space for the moral considerability of non-humans and a recognition that human life (including the 'good life') does not require the insensitive or inhumane treatment of those with whom we share the planet. As Watson assets:

> It is nice that human survival is compatible with the preservation of a rich planetary ecology, but I think it a mistake to try and cover up the fact that human survival and the good life for man *is* some part of what we are interested in. (1983: 256)

This is precisely the point of the ethics of use position. It is not against human interests to acknowledge moral considerations in respect to our dealings with nature. This recognition does not preclude our purposive use of it. In terms of practical outcome, there is an increasing convergence between human interests and non-human welfare, particularly in respect to care for descendants and future generations. Policies based on ecocentric or animal rights moral arguments often lead to similar treatment of non-humans as those grounded in human interests. This is especially so in regard to future human generations, where we can say that our continued flourishing (both as a species and as a collection of particular cultural groupings) demands the type of habitat and ecosystem protection that would result from a concern with non-human welfare.

In rejecting the notion that green normative arguments must be understood as the proposition of a 'new ethic', and insisting that the spheres of human–nature and intrahuman action must be kept separate, I conclude that there *are* resources within the Western moral tradition, and within existing Western culture, that greens can tap into and use. Demands for a 'new ethic' because the existing one is 'speciesist' are misplaced. Conventional moral reasoning can, like Achilles's lance, heal the ecological wounds it has inflicted. According to Holland:

> opponents of speciesism are right about the shabby treatment of other animals meted out by humans, and also about the need for a radical reformation of human attitudes. But I am not convinced that speciesism is the villain of the piece, and am more inclined to suspect some of the more old-fashioned vices such as cruelty, lack of sensitivity and lack of understanding. (1984: 291)

And to combat these and other ever present ecological vices we need ecological virtues. In the end then what green politics comes down to is the claim that the normative roots of our ecological crisis lie in the rather familiar moral territory of greed, short-sightedness, intemperance and ignorance. As Passmore remarks, 'What it [the West] needs, for the most part, is not so much a "new ethic" as a more general adherence to a perfectly familiar ethic' (1980: 187).

Aquinas wisely pointed out that it is better for a blind horse to be slow. Likewise it would be better for contemporary society, given the uncertainty that marks our dealings with the environment, coupled with the technological capacity to alter the conditions of life on earth as we know it, to moderate its demands for material consumption (ends), and adopt a more prudent disposition in the employment of its means. After all, we do not have exclusive ownership of the earth but share it with other species and future generations of humans, and while we transform it to fulfil our interests, we must be ever wary that we do not unjustifiably harm the interests of other species or those who will come after us. The earth was not made for us, but neither were we made for the earth. Finding a mean between these extremes of 'arrogant humanism' on the one hand and 'ecological quietism' on the other is, as I hope to have shown in this and the previous chapter, the goal of green politics as ecological stewardship.

Having outlined what I offered as a defensible anthropocentric moral basis for green politics, I shall in Chapters 4 to 7 seek to develop a conception of green politics consistent with that ethical basis. However, before moving on to outlining this alternative conception, the next chapter takes a critical look at eco-anarchism, which in many ways stands to green political theory in much the same way as deep ecology does to green moral theory.

Notes

1 As will be made clear later on, distinguishing between the concept of anthropocentrism and particular conceptions opens up the possibility that some conceptions of anthropocentrism are compatible with green moral claims concerning the treatment of the non-human world. What critics of anthropocentrism need to demonstrate is that this type of moral reasoning *necessarily* leads to the devaluation of nature in theory and its mistreatment in practice. Following on from the argument in the last chapter, the position adopted in this chapter is that there is no necessary link between anthropocentrism and the devaluation or mistreatment of nature.

2 That human–nature relations always had an ethical dimension is a theme developed by many writers in what I call the 'naturalistic tradition' such as Midgley (1983a; 1995) and Benton (1993) whose work I draw upon in this chapter. For some contemporary evidence of this feature of human dealings with nature see O'Neill, who argues that the high number of 'protest bids' (where individuals place an extremely high value on preservation of some part of the environment), recorded during environmental cost–benefit analyses, demonstrates 'ethical commitment' (1993: 120). See also Brennan (1992: 17). Other more intuitive grounds for the ethical framing of human–nature relations is given by the 'last-person argument' which demonstrates that, faced with the hypothetical situation of being the last person on earth (Lee, 1993b) or the only human on a deserted island (Midgley, 1983b), most people feel that the destruction of the earth or the island would be morally wrong. However, the hypothetical and abstract nature of such reasoning is often of little help in assessing actual, concrete human–nature relations which are rarely related to such exotic examples. Later, the moral censure attached to forms of human use of the world characterized as 'wanton', 'unnecessary' or not fulfilling a serious human interest is used to flesh out the reflexive nature of weak anthropocentrism. This is held to consist in holding that human interests are necessary but not sufficient to justify particular human uses of the non-human world.

3 Generally speaking, something like a 'great chain of being' (Pepper, 1984) perspective is usually (albeit tacitly) accepted as, *ceteris paribus*, the most appropriate moral framework within which *actual* human–nature relations can and ought to be judged.

4 Other examples of this discrepancy include the premium attached to 'naturalness' and 'natural' (from advertising to ordinary discourse), at a time when nature itself is under threat.
5 It may be worthwhile indicating here that although the ethical position defended here seems less radical than standard non-anthropocentric positions, this does not imply that changes in the treatment of nature sanctioned by this immanent critique may not require large-scale, and sometimes radical, alteration of human social practices. Ethical 'reformism' is perfectly compatible with political radicalism.
6 Sympathy is also an ecological virtue, as distinct from the 'vice' of sentimentality, as suggested in the last chapter. As Gruen notes, 'sympathy is fundamental to moral theory because it helps to determine who the proper recipients of moral care are' (1993: 351).
7 This conflation is less marked within one strand of green moral theory, the 'animal liberation' literature, where there is greater sensitivity to the non-rational (which is not to say irrational) dimensions of morality. This strand does not fall into the deontological animal rights approach (Regan, 1983) or Singer's (1990) utilitarian theory, but emphasizes the role of sympathy and actual contact between humans and animals. This includes Midgley (1983a; 1992), Clark (1982) and others in what can be called the 'Humean tradition', and more recent manifestations in the form of extending the feminist 'ethic of care' argument to the treatment of animals (Gruen, 1993; Plumwood, 1993).
8 A full account of a Humean-inspired naturalism would have to include a critique of the pre-eminence of 'human autonomy' and 'voluntarism' within Kantian moral theory, from the point of view of the significant part played by those non-chosen and given aspects of human nature and human relations from which the former abstracts, yet which play a pivotal part in human experience and relations. Our neediness, vulnerability and dependence on others are something that has only recently been suggested as at the heart of moral life by feminist moral theorists, amongst others. A naturalistic view also shares the feminist concern with 'embodying' autonomy, as well as opening the way toward modifying the view that moral duties are paradigmatically 'chosen' or arise from previous voluntary acts as suggested by Hart (1955). See O'Neill (1991: 298–9) on the moral significance of human neediness, and Goodin (1985) for an attempt to develop a theory of morality based on the centrality of 'vulnerability'.
9 For anthropological accounts of the various cultural and mythical meanings attributed to such categories, particularly the notion of 'wilderness', see Rennie-Short (1991). What I intend to convey by the qualifier 'in less enlightened times' is the sense that the usual negative moral resonance associated with the second set of terms is difficult to sustain in the contemporary world. On the one hand the 'disenchantment of the world' has largely drained such terms of their negative and often threatening character, while on the other hand, ecological science has shown us that what initially looks like a 'barren wasteland' is in fact a rich ecosystem supporting a multitude of life forms. To think of it as 'barren' (valueless) is a mistaken judgement, based on ignorance, and therefore is an example of an ecological vice. The link between ecological science and environmental virtue is discussed later.
10 The 'otherness' of nature as the appropriate perspective within which to place human–nature relations is I think correct, not least because it follows

from the naturalistic position outlined here, but also because it tempers one aspect of the 'arrogance of humanism' concerning the ascribed conviction of the latter in the ultimate transparency of the natural world to human reason. On the point about limits to human knowledge of the world see Hayward (1995). As O'Neill notes in reference to the centrality of science in demonstrating the indifference and strangeness of nature, 'The assumption that the discovery of nature's impersonality and strangeness is something to be regretted, a cause of the "disenchantment of the world", needs to be rejected' (1993: 151).

11 In many respects the distinctions drawn between use/abuse and use/non-use are similar to Kantian concerns relating to treating humans always as 'ends in themselves' and never purely as means. The Kantian injunction is not that we ought never to treat humans (or at least rational ones) as means, but rather that we should always regard them as ends in themselves as well. This implies that any 'use' of one human by another (for example, employment) ought to be tempered by moral considerations not related to that use. In a comparable fashion, an ethics of use seeks to persuade us that some human uses of nature ought to be governed by wider human interests which temper narrow considerations of economic efficiency, for example.

12 This raises an issue for any suggested 'green theory of justice' in that it would have to balance a concern for cultural uniqueness and difference (which is partly constituted by particular uses of the non-human world) and upholding ecological principles which may come into conflict with these culturally specific uses of nature. Thus, it is possible to see in the 'ethics of use' argument some indication for any green theory of justice in respect to the likely distributive conflicts as a result of the implementation of green policies.

13 The question we must ask is whether or not deep ecology associates green politics with a view of the good that unnecessarily constrains it in terms of the balance between the right and the good in its political principles. Is the deep ecology vision simply too tightly tied to particular social practices that it limits the range of positions, lifestyles, and uses of nature that can be described as 'green'? How tolerant in other words would a green politics premised on the deep view of the 'good life' be of other views of the good? Just as Plato banished the artists, would green politics banish the economists or 'consumers'?

14 'Symbiosis' and 'parasitism' convey the general moral nature of social–environmental relations, while sustainability and unsustainability convey the productive relations between society and environment. There is no assumed 'mapping' of these two sets of criteria such that morally symbiotic social–environmental relations are also sustainable and vice versa. It is perfectly possible for a society to be morally parasitic and ecologically sustainable.

15 This discrimination between consumer and citizen is common in green politics, and attention to the relationships and possible reconciliation between them have increasingly occupied centre-stage in debates. The most obvious example of this is the 'sustainable development' debate which, at root, can be seen as an attempt to reconcile the interests of consumers in economic growth with their interests as concerned 'green citizens' in ecological sustainability. Sustainable development is thus, in part, a deeply normative issue in which culturally, that is locally defined, values and

relationships, economic on the one hand and normative-environmental on the other, can, in principle, be reconciled. See Sagoff (1988), Brennan (1992) and Keat (1994).

16 A deeper understanding of this citizen/consumer distinction can be found in Arendt. Anticipating green concerns, Arendt argued that the growth of appetite, expressed as exponential economic growth within a consumer-driven economy, 'harbours the grave danger that eventually no object of the world will be safe from consumption and annihilation through consumption' (1959: 115). Here citizenship, while obviously premised on human consumption, sets parameters on the latter in the interests of realizing the ideals of the *vita activa* or human action.

17 Having different ecological virtues and proper modes of relating to different parts of the non-human world is vital, particularly given the increasingly distant and mediated experience people have of nature in a rapidly urbanizing world. Hence the important political point of finding ecological virtues and modes suited to the urban character of modern human life. All of which does not preclude argument for protecting wilderness and rural areas from urbanization.

18 This is similar to Hayward's defence of 'humanism' understood as enlightened self-interest, which he claims 'amounts to an assumption of an underlying rational harmony between the interests of human and non-human beings' (1995: 63). However, as indicated in Chapter 5, this harmony is dynamic not static and has to be collectively created and maintained, rather than 'given' or 'self-evident'.

4

Rethinking Green Politics and the State: A Reconstructive Critique of Eco-anarchism

CONTENTS

The characterization of green politics as either 'anarchistic' (Dobson, 1990: 83–4; Freeden, 1995: 16–18) or a modern form of anarchism (Bookchin, 1971; 1992a) has wide currency. Both within and without the green movement, its distinctiveness is held to reside in its embodiment of traditional anarchist values expressed within a contemporary ecological political idiom. This self-understanding is particularly evident in the almost complete monopolization of the green political imagination by an anarchist vision of the society greens would like to create. The many pastoral utopias that litter green political literature pay eloquent testimony to the common view that 'greens are basically libertarians-cum-anarchists' (Goodin, 1992: 152). While 'soviets plus electrification' equalled socialism for Lenin, it seems that, for many green theorists, activists and commentators, stateless, self-governing communities plus solar power equal the 'sustainable society'. Although 'small is beautiful', beauty is often in the eye of the beholder. And in the case of the eco-anarchist vision, the disproportionate attention given to describing the future society often means that less attention is given to the necessity of developing a critical analysis of the present situation, and of paying attention to the problems of political transformation, that is, getting from the unecological present to an ecological future. The aim of this chapter is not to banish the eco-anarchist vision from the green pantheon, but rather to critically interrogate it with a view to integrating its insights within the alternative conception of green political theory being developed here.

Several reasons can be given for the negative attitude of greens to the state. Firstly, the strong influence of anarchism on the early development of green political theory underwrites a rejection of the state on the grounds that it is inextricably bound up with the ecological, political, social and ethical problems that greens are concerned with solving. Thus the state's continued existence is held to exacerbate the fundamental, underlying cause of which the ecological crisis is an effect. The state is regarded as protecting an 'environmentally hazardous dynamic' (Carter, 1993: 45–9), or the embodiment of 'materialism, institutionalized violence, centralization, hierarchy', values and practices antithetical to the green perspective (Porritt, 1984: 216–17).

Secondly, a suspicion of approaches that give a prominent place to formal political and legal institutions is tied up with the authoritarian nature of previous green defences of the state, which stand at odds with its self-professed libertarian and democratic character. The undemocratic character of 'realist' solutions to the ecological crisis, as in the authoritarian proposals of Ophuls (1977), Heilbroner (1980) and Hardin (1977), has made greens weary of state-centred approaches to environmental problems.

Thirdly, more 'benign' readings of the liberal state's role in dealing with ecological problems are held to invariably favour 'technocratic' and administrative, as opposed to 'democratic', solutions. In other words, only a 'technocentric' outlook (O'Riordan, 1981: 11–19), premised on an instrumentalist valuation of nature and a reductionist problem-solving methodology, is compatible with the bureaucratic-administrative logic of state agencies. Common to these arguments against the state is that talk of the state having a positive role in green theory and practice belies an 'environmental' rather than an 'ecological' perspective (Bookchin, 1980: 70; Dobson, 1990: 13). That is, talk of a 'green state' is an oxymoron at best, or a betrayal of the radical promise of green politics at worst. The strength of feeling generated by the place of the state within green politics can be most dramatically seen in the ideological division within the green movement between 'fundi' (radical) and 'realo' (pragmatic) elements (Doherty, 1992). This division parallels, and is related to, the distinction between 'ecological' and 'environmental' perspectives on green ideology. Environmentalism is typically presented as a reformist strategy whose principal focus is the 'greening' of contemporary liberal democracies, particularly the instrumental use of nature, rather than the seeking of widespread social and economic change to create a particular vision or blueprint of the 'sustainable society', which is the aim of 'ecologism' (Dobson, 1990: 73–130). For present purposes this distinction between environmentalism and ecologism will be assumed to centre mainly around opposing attitudes to the state, with ecologism being unequivocally anti-state and environmentalism more agnostic.

Fourthly, and following on from the above, proposing state-centred solutions to the ecological crisis is 'reformist' not only in political but also in ethical terms. Articulations of the ecological crisis which give prominence to the state 'crowd out' the need for radical moral change. In other words, the technocentric outlook which favours reformist as opposed to more radical institutional change also underwrites a continuation of the anthropocentric moral outlook in general and the treatment of nature as a stock of raw materials in particular.

Two anti-state eco-anarchist theories in particular will be examined in this chapter; bioregionalism and social ecology. After critically analysing these two theories and finding them wanting in their rejection of the state, I then move on to argue that the *transformation* rather than the *abolition* of the state and civil society is closer to green political aims and values. The argument of this chapter is that as one moves along the green political continuum from bioregionalism to social ecology, the eco-anarchist position 'shades into' an understanding of green politics which sees its primary goal as the *democratic transformation* of the state and civil society. This chapter seeks to establish a role for the state within green political theory by utilizing the analytic distinction between the state and civil society, which is further developed in Chapter 7. At the same time it is also hoped that this critical analysis may integrate eco-anarchist insights within green political theory. This sees the eco-anarchist contribution to green politics as resting in *honouring* as opposed to *realizing* eco-anarchist values. The chapter's overall aim is to forge a workable synthesis between the anti-state and pro-state positions within green political theory.

Eco-anarchism: Strong and Weak Versions

What O'Riordan has called the 'anarchist solution' (1981: 307) has been an enduring part of the green ideological spectrum. It is not difficult to see why anarchist forms of social organization have appealed to greens. Anarchism argues that if left to themselves human beings will naturally cohere into 'organic' communities regulated by principles of mutual aid and sociality. As such it coheres well with a 'naive naturalism' that peppers the literature on green politics: the idea that we can 'read off' social relations from nature. The quality of social relations within stateless communities is such that the laws, procedures and institutions of the state are unnecessary for governance. In short, anarchism challenges the traditional defence of the state, that it alone can provide 'public goods', particularly social order and environmental quality.

Green political theory is usually thought of as representing an inherently anti-state position. Most green writers and commentators seem to agree that its general anarchistic complexion is one of its unifying features. Green theorists as different as Naess and Bookchin, for example, can both agree on the basic stateless nature of the type of society they envisage. Naess can say that 'supporters of the deep ecology movement seem to move more in the direction of non-violent anarchism than toward communism' (1989: 156); while Bookchin claims that his social ecology theory is premised on the abolition of the state (1992b: 95–6). That a commitment to this 'anarchist solution' may actually undermine crucial aspects of green political theory is rarely fully addressed, although some green theorists and commentators are aware of the problems raised by green anti-state arguments (Dobson, 1990: 183–6; Eckersley, 1992a: 175, 181–6; Young, 1992).

One prominent form of eco-anarchism, bioregionalism, begins from the argument that the resolution of the ecological crisis calls for greater integration of human communities with their immediate environment, with 'natural', rather than 'human-political' (state), boundaries delimiting the appropriate human social unit. This claim that the state is unnecessary for securing and enjoying public goods I call the 'weak' version of eco-anarchism, to distinguish it from the stronger claims made by social ecology. While bioregionalism envisages an ecological stateless society, roughly along the lines of the traditional anarchist vision of a 'commune of communes', it does not base its claims on a critical assessment of the state in political theory and practice. Social ecology, on the other hand, does.

For social ecologists, 'The state consolidates and protects the family of hierarchies [class, gender, race, age, mind–body] becoming a hierarchy in its own right' (Kossoff, 1992: 8). Carter (1993) develops a comprehensive green anarchist theory, which highlights many of the concerns of social ecology. Holding a view of the state as 'an autonomous agent' (1993: 45), he claims the state can never be used to serve civil society (1993: 42). This discounting of the dominant 'instrumentalist' view of the state, that is, seeing the state in functional terms as an 'enabler' to civil society, is the hallmark of what I want to call the 'pure' anarchist position. On a pure anarchist reading the state has its own interests and agenda, and its sole *raison d'être* is the systematic exploitation of society as a whole. Adding a feminist perspective, Bookchin holds that 'The institutionalized apex of male civilization was the state' (1990: 66). Social ecology advances what I call a 'strong' version of eco-anarchism. It goes beyond claiming that the state is *unnecessary* for ensuring the public goods of social order and environmental protection, to stipulating that it is positively *undesirable.* It is undesirable because by its very nature the state is a coercive institution which curbs human freedom. Thus from the strong or pure anarchist position, the state is not just part of the ecological problem; it *is* the problem. It is the Gordian knot, the severing of which is a necessary condition for the creation of an emancipated, ecologically rational society.

Central to both the strong and weak versions of eco-anarchism then is the conviction that only the transcendence of state institutions, and their replacement with informal, community-based social mechanisms, will produce the social conditions for the realization of green aims. The argument for stateless self-determination is summed up by Taylor as resting on a version of communitarianism: 'If we want to do without the state or substantially reduce its role, we have to revive and rebuild communities' (1982: 4). This distinction between strong and weak versions of eco-anarchism is significant. The strong version of anarchism is problematic, particularly as regards the lack of empirical evidence to support its claims, while theoretically it can be criticized as dependent upon an ahistorical explanatory schema which confuses the *concept* of the state with particular *conceptions*. The pure anarchist position in which the state is an intrinsically exploitative institution is, I suggest, both unnecessary and unhelpful to green politics. At the same time the traditional anarchist demand that social relations be transparent is neither a necessary nor a desirable political aim of green politics, which is not to say that all anarchist principles or values are to be rejected.[1] Indeed, as argued below, a close analysis of the social ecology position is compatible with the democratization and decentralization rather than the abolition of the state.

Bioregionalism

Bioregionalists place a premium on the necessity of strong, affective senses of community and communal identity, and see the ecological problems we face as due in no small part to the demise of community. Although this decline in community is a common observation of communitarian critiques of 'modernity' and the process of 'modernization', bioregionalists locate one of the roots of its decline in the disengagement of people from a specific land base and rural way of life or modes. With no enduring link to the land, community is 'rootless' and individuals are vulnerable to the anomie and alienation of 'mass society'. The movement of people from the land into the cities, which is one of the chief characteristics of the process of industrial-based modernization, also marks a decisive increase in ecological degradation and a shift in ecological consciousness. On the one hand, with fewer people in agriculture, there is a shift towards the capitalization of the agricultural economy, with pesticides, heavy machinery etc. increasing the (short-lived) productivity of the land. This process leads to ecological degradation as more and more pesticides, fossil fuel etc. have to be used to keep the land fertile and productive. This was discussed in Chapter 2 in terms of the shift from agriculture as a social practice and way of life to an industrial business, namely agri-business. On the other hand, the demographic shift from a rural, agricultural lifestyle to an urban, industrial one means that for most people the 'environment' becomes the human, built one, while the non-human environment is simply not part of their lived experience, even though they still depend on it. Thus there is an increasing gap (both actual and cognitive) between the environment as a sphere of human production and the environment as a resource for human consumption, as a result of altered social–environmental relations.

People in modern society according to bioregionalism have 'forgotten' that the economy and all its works is a subset of and dependent upon the wider ecosystem. Bioregionalists argue that whereas it may be possible to have a 'post-industrial society' we cannot have a 'post-agricultural society'. Yet this is precisely the 'misperception' of the majority of citizens in the Western world. Modern citizens have not only lost contact with the land, and their sense of embeddedness in the land, but at the same time they have lost those elemental social forms of more or less intimate and relatively transparent social relations. Thus a basic aim of bioregionalism is to get people back in touch with the land, and constitutive of that process is the re-creation of community in a strong sense. This sense of community is close to the *Gemeinschaft* view of community, in opposition to a looser sense of *Gesellschaft* or 'association'. Here bioregionalism is close to deep ecology, indeed it would be fair to say that bioregionalism is the politics of deep ecology. For example, according to Sessions, 'Many supporters of the Deep ecology movement believe that human habitation on Earth, including the cities, should ultimately be bioregional' (1995a: 416). Both share the belief that only a radical bioregional transformation of society, at the economic, social, political and cultural levels, can resolve the 'total crisis' of contemporary societies as discussed in Chapter 2.

This 'return to the land' ethos is eloquently expressed by D.H. Lawrence:

> We must get back into relation, vivid and nourishing relation to the cosmos and the universe . . . Vitally, the human race is dying. It is like a great uprooted tree with its roots in the air. We must plant ourselves again in the universe. (1968: 410)

Bioregionalism sees itself as offering a way in which we can return to the 'land' and 'replant' ourselves again in the natural world. In the words of Kirkpatrick Sale, we must become 'dwellers in the land' (1984a: 224) if we are to comprehensively resolve the ecological crisis.

Gemeinschaft and Gesellschaft

For the nineteenth century sociologist Tönnies, the emergence of the modern social world was marked by a movement from 'community' or *Gemeinschaft* to a looser sense of 'society/association' or *Gesellschaft* (1957: 33). Instead of individuals collectively regulating themselves through various traditional community structures, (external) institutions (state and market) became necessary to 'order' an emerging 'civil society' of individual competitive wills. From a bioregional point of view this movement also accounts for the ecological crisis. The emergence of civil society in the modern age, particularly the movement from the land to the city, the creation of extensive trade and other relations which went beyond the local or national hinterland, as well as the evolution of 'mass society', have all contributed to severing the link between human community and the land, and rendered opaque and mediated many human social relations (including those with nature). Hence its concern with re-creating *gemeinschaftliche* communities along ecological or bioregional lines. This is well expressed by Jones who declares: 'If a future society based on the Gaian principles of inter-dependence, mutuality and inter-relatedness is to be achieved, a re-emergence of some form of *Gemeinschaft* is essential' (1990: 109).

One of the earliest and most consistent green proponents of strong communities is Goldsmith who as long ago as 1972 declared:

> cultural and economic heterogeneity is associated with a state-like political structure . . . Only an elaborate bureaucracy run by a shameless autocrat can hope to control a mass of people deprived of a common culture and a sense of duty to their society (Goldsmith, 1972a: 253)

The corollary is clear: if we seek to do without state institutions, forms of social life with a 'common culture' and a 'sense of duty' derived from that culture are the only alternative. Since the state is a historical, contingent/artificial, social form, there is nothing inevitable about the continuation of state-centred societies. Therefore, non-state forms of social organization are always possible. The logic of Goldsmith's position is that stateless social order not only requires the re-creation of community, but also requires a particular type of community, a prominent characteristic of which is a low degree of internal plurality.

Taylor (1982) spells out why anarchism demands communities of this type. For him, stateless social order requires that the community have (a) strongly shared beliefs, (b) relations between members that are direct and many-sided, and (c) social interaction that is characterized by reciprocity and mutual aid (1982: 26–30). It is the *quality* of relations within such communities that allows the operation of stateless mechanisms to get people to do things they would not voluntarily do, yet are necessary for social order and other public goods. This addresses two common misunderstandings of anarchism, namely that it implies social 'chaos' and relies on a 'myth of collective harmony'. The former misunderstanding is typically expressed in the pejorative connotations of the term 'anarchy', a term which owes much to the Hobbesian conception of the 'state of nature' as the description of stateless disorder. Anarchism, as opposed to the pejorative term 'anarchy', does not imply social chaos and it explicitly rejects the Hobbesian vision of what social life is like without the state. To those who claim that the selfishness or 'badness' of human nature undermines the anarchist case for voluntary social co-operation, anarchists reply:

> The assumptions made by Hobbes and Hume were supposed to characterise human behaviour in the absence of the state; but perhaps they more accurately describe what human behaviour would be like immediately after the state has been removed from a *society whose members had for a long time lived under states*. (Taylor, 1976: 141, emphasis added)

In other words, with the decline of 'community' and the rise of the state, individuals have lost the capacity to govern themselves without the state.[2] As Taylor argues, 'the state is like an addictive drug: the more we have of it, the more we "need" it and the more we come to depend on it' (1976: 134). However, in opposition to this extreme view, one may view the state instrumentally, as a prosthesis rather than as an addictive and dangerous drug.

The second misunderstanding is the assumption that anarchism posits that once the state has been abolished, individuals will 'automatically' or 'naturally' be in harmony with one another. This 'myth of social harmony' is rejected by realist anarchist theorists such as Taylor and eco-anarchists such as Goldsmith and bioregionalists such as Sale. All these anarchist theorists agree that the provision of the public goods such as social order and environmental protection depends on getting people to do things they might not want to do. Hence 'realistic' anarchists argue for the abolition not of social coercion, but rather of one particular form of it, namely the forms of institutionalized coercion employed by the state. Socialization, the internalization of communal norms and conventions, together with forms of social sanction such as public ridicule, stoning and shaming, perform the necessary coercive functions within stateless communities (Taylor, 1982: 80–7; Goldsmith, 1972b). Thus the prevention of 'free-riding', a precondition for the provision of collective goods, can be achieved without recourse to the state, but not without the employment of social, non-state forms of coercion. The state may be abolished within an anarchist society but coercion and power are not.

A large part of the process of recovering such communal capacities from a bioregional point of view requires 'reinhabitation', the conscious reintegration of human communities within their local bioregion (Eckersley, 1992a: 167). This is not just on the grounds that small-scale communities dependent upon and living in close contact with 'their' local environment are less ecologically destructive than large societies. Rather, reinhabitation, becoming 'native' to a place, is also held to be an identity-constituting ecological condition. *Who* you are is a question of *where* you are, the types and quality of relations you find yourself in, both social and ecological. Reinhabitation of the land is necessary not only on ecological grounds but also on moral and political grounds. Part of the latter involves the creation of a bioregional community, a central aspect of which is knowing one's bioregion through experience, as a way of becoming 'rooted' once again in the land.

The basic bioregional vision is of a patchwork of self-sufficient, small-scale, ecologically harmonious communities, organized according to their own normative standards (Sale, 1984a: 233). Bioregionalists (like many greens) discourage trade and stress the ecological and social benefits of communal autarky. One of the main reasons given for this is to encourage people to live within the limits set by their local environment rather than to depend on distant ecosystems or the planet as a whole for resources, goods and services. Dasmann's distinction between 'ecosystem people' and 'biosphere people' captures the basic bioregional position. According to him:

> Biosphere people draw their support, not from the resources of any one ecosystem, but from the biosphere . . . [They] can exert incredible pressure upon an ecosystem they wish to exploit . . . something that would be impossible or unthinkable for people who were dependent upon that particular ecosystem. (in Sessions, 1993: 121)

Each ecosystem is unique, and thus demands a particular way of life. In this way the community's distinctiveness is intimately related to how it interacts with its local ecosystem or bioregion. This interaction, the metabolism between community and environment, and thus the community's identity, is co-determined by the bioregional ecological context. Defending biodiversity and cultural diversity thus go hand in hand in the bioregional scheme of things, particularly in respect to wilderness conservation and indigenous people as 'authentic inhabitants' or 'ecological guardians' of wilderness areas.

From the bioregional position, it is trade between disparate parts of the world as a result of the global market that has created the impression, particularly in those Western nations who benefit most from such planetary-wide exchange, that escaping the confines of their particular ecological context is proof of transcending 'natural limits' as a whole.[3] This is the ecological slant on the historical necessity for the imperial expansion of Western societies, and the ecological reason for present day neocolonial trade practices, which from the green perspective underwrite current North–South inequalities. Simply put, countries such as Britain, France and the United States could never have achieved the rates of material economic growth and affluence they continue to enjoy if they had been dependent upon their own native natural resources. The affluent lifestyle enjoyed in these countries is dependent on a disproportionate consumption of world resources. To many greens these economies are, in strict ecological terms, unsustainable, in the sense that their models and modes of development are physically impossible to achieve on a global scale (Whiteside, 1994: 340; Goodland, 1995). Bioregionalists encourage economic self-sufficiency which, by rendering the community dependent upon its local ecosystem, encourages prudence and the adoption of a long-run perspective. One way of looking at this is to say that a bioregional view encourages us to see the productivity of the earth in terms of a renewable (but limited) flow of 'income' rather than as 'capital' (Daly, 1973a). In this way bioregionalism proposes to save the whole (the global biosphere) by saving the parts (individual ecosystems), as well as fostering the independence and cultural uniqueness of bioregional communities.

There are several positive lessons which can be drawn from bioregionalism. These include the emphases on economic/ecological independence, on ecosystem sensitivity, and on the environmental destruction and social exploitation attendant upon global trade. However, the bioregional vision is also flawed in some significant ways. The first refers to issues arising from the lack of interaction between bioregions in the context of the distribution of resources across the face of the planet. Simply put, the autarky imperative coupled with strict ecosystem dependence implies that those living in resource-poor ecosystems are condemned to their fate as there is no provision for the redistribution of resources between bioregions. The redistribution of resources across the planet, regarded by many as a core part of any green or environmentally informed theory of global distributive justice (Attfield and Wilkins, 1992; Pasek, 1994), seems to go against the communitarianism which underwrites much of bioregionalism. Transfers, whether from trade or charity, may compromise the distinctiveness of bioregional communities, since their identity as a community is tied up with how they live within and use 'their' ecosystem, as opposed to those of others. The wholesale global redistribution of resources is based on prioritising a universal 'biosphere/human-species' relationship over any particular 'ecosystem culture', a view that sees the biosphere rather than the ecosystem as the 'resource base' for human welfare.

At the same time redistribution on any large scale, like trade, seems to be ruled out as a homogenizing process, destructive of cultural difference and diversity. According to Berg, 'Global Monoculture dictates English lawns in the desert, orange juice in Siberia and hamburgers in

New Delhi. It overwhelms local cultures and "raises" them regardless of the effects on cultural coherency or capacities of local natural systems' (1981: 25). Even 'cultural' exchange, expressed in such practices as tourism, even if ecologically sensitive, is discouraged as being destructive of rootedness and communal distinctiveness (Mills, 1981: 5). As such bioregionalism is arguably the most communitarian strand within contemporary communitarian thought. An extreme interpretation would be that resource-poor bioregions and communities simply have to survive and flourish as best they can on their own without any (or many) external transfers.

The bioregional vision of a world made up of self-sufficient, ecologically harmonious bioregions harks back to a pre-modern era before exploration, trade and cultural exchanges brought people from different parts of the world together and gave tangible expression to the idea of the 'human species'. In place of the global village, with its communications networks, global political and economic institutions (*inter alia*, a world market, the World Bank and the United Nations), bioregionalism implies a 'refeudalization' or 'Balkanization' of the world into ecologically defined political and economic units (Sale, 1984b: 171).

Other problems with bioregionalism can be grouped under two headings: those relating to internal relations within, and those relating to external relations between, bioregional communities. On the former, Dobson (1990: 122) points out that there is no guarantee that bioregional communities will be democratic or just. Indeed, according to Sale, 'truly autonomous bioregions will likely go their own separate ways' (1984b: 170). The reason for this is that bioregionalists place the communal right to self-legislate as the highest social value. The affirmation of communal solidarity is prioritized over 'contingent' and non-local values such as equality, fairness or democracy. This communitarianism, together with the assumption that social and ecological problems are rooted in 'bigness' (Sale, 1980: 82) or a lack of appropriate or 'human' scale, are the two principles around which bioregionalism is woven. However, small is not always beautiful, and small scale, although an important consideration, is not a panacea for all social and ecological ills.

An examination of how bioregionalism copes with internal differences brings into sharp relief its problematic place in green political theory. Conflict within bioregional communities, according to Sale (1980), should not require recourse to formal principles of justice or political institutions external to the community. For example, the 'natural' way to deal with disputes between an aggrieved minority and an implacable majority is for the community to divide, with the minority to settle elsewhere. From the bioregional point of view, 'The commodious solution is not *minority rights but minority settlements*' (1980: 480, emphasis added). This 'fissioning' of communities when they get too large or develop tensions incompatible with communal consensus is, according to Taylor, 'a normal part of the life of stateless societies' (1982: 92; Sale, 1984b: 170). But in a non-Lockean world, that is where there is no unsettled territory, this bioregional 'solution' to serious conflicts within most societies is simply unworkable. In a closed world, there is no 'away' to which the displaced can go without entering the territorial jurisdiction of another political community, unless the existing territory is divided between the two sides. This recourse to fissioning and relocating is surprising given the strong link made between communal and personal identity and territory. Perhaps the bioregional point is that people must make a choice between exile (and thus compromising their identity) and putting up with the discomfort of being in serious disagreement with the rest of society. In other words, individuals are required to rank community membership relative to other values. If this is the case then bioregional communitarianism takes the slogan 'My country right or wrong' extremely seriously. However, one has only to look at the conflict within such divided societies as Northern Ireland, the former Yugoslavia and the former USSR to see that 'minority settlements' do not constitute a realistic, never mind a just, solution. Whether eco-anarchists like it or not, the history of the nation-state (and not necessarily the liberal-constitutional version either) provides ample evidence that it can protect the rights of minorities and individuals as much as it can hinder them. The instrumental view of the state, namely that it can serve the interests of civil society, *contra* the strong anarchist thesis, is not completely mistaken.

This 'justice as displacement' argument is premised on protecting the community's sense of identity and solidarity from those who argue for a different understanding of communal identity, shared goods, history or meanings. Hence the distrust of appeals to justice as an entrenched system of individual rights and liberties which transcend local norms. Justice as an ethical perspective that transcends communal conventions is either incompatible with complete communal autonomy, or is unnecessary for social order. Like other communitarians, bioregionalists (and some social ecologists) regard justice as a remedial virtue, useful for remedying flaws in social life (Kymlicka, 1993: 367): flaws that are the result of a decline in community. That bioregional communities would tend to be conservative is accepted by bioregionalists as the necessary price to maintain what Kumar (1978) has called the 'intimacy of scale', necessary for social order within stateless societies. According to Sale, 'To be sure . . . consensual communities will tend towards conservatism . . . but it will by the same token make them more stable, more predictable, more "comfortable", and less prone to ill-considered decisions' (1980: 501). However, one does not need consensus based on tradition in order to make collectively prudent decisions.

The danger here is not just the threat to individual liberty within extreme forms of communitarianism, where the individual as an 'organic part' of the wider collectivity can have her interests sacrificed for the benefit of the 'common good'. Rather the conservative possibility inherent in aspects of the green position illuminates the threat to plurality and social diversity and thus the precondition for democratic politics. A strong critique of bioregionalism would be that social and ecological harmony for it is 'natural', i.e. 'given', making the issue of social and ecological order in some ways 'pre-political'. A clear example of the conservatism expressed in some versions of bioregionalism can be seen in the emphasis placed on communal self-identity being constituted by a religious or quasi-religious outlook. In bioregionalism this outlook is provided by deep ecology and the 'reinhabitation' process, and has affinities with the 'eco-monastic' strategy associated with Rudolf Bahro (1994; Eckersley, 1992a: 163–7). In both bioregionalism and eco-monasticism, it is a strong possibility that the community is held together by a shared metaphysical view (usually spiritual or mystical) of the natural world in general and the local ecosystem in particular, and the community's relationship to them (Sale, 1984a). Bioregionalism seems to endorse cultural diversity between bioregional communities and favours homogeneity within them.

As argued earlier in Chapter 2, it is extremely unlikely that a spiritualized view of the natural world (as opposed to a moral view informed by science) would succeed in 'converting' Western citizens to green principles. More damaging to green theory in terms of its 'progressive' self-understanding is that societies infused with a strong shared religious sense have typically functioned as fertile breeding grounds for intolerance. A possible response to this is that the spiritualized worldview put forward by bioregionalists and deep ecologists is inclusionary, welcoming difference and otherness. However, it remains to be seen if such flexible, non-dogmatic, shared spiritual visions are sufficiently robust to furnish communities with strong senses of shared identity in the sense required for stateless social mechanisms to work. Simply put, there is every reason to believe that the tolerance proclaimed for green metaphysics and bioregionalism may be undermined by the emphasis on 'tribalism' (Sale, 1984b), and the requisite degree of social homogeneity that is required for stateless social order.

It is perhaps at the external level that the shortcomings of bioregionalism are most apparent. Given the ongoing impact of globalization on human societies, drawing them into an increasingly complex web of interrelations, the complete realization of the bioregional vision, complete bioregional autarky, is impossible. Which is not to say 'deglobalization' may not be necessary or desirable from a green political point of view. When we look at the global nature of ecological problems such as ozone depletion and global warming, there is a need for more not less co-operation and interaction between societies. From a global ecological point of view, the fragmentation of the world as propounded by bioregionalism may exacerbate ecosystemic problems. The strategy of saving the whole by saving the parts only works if there is some degree of trans-communal co-operation and co-ordination. This is because when it comes to ecosystems, Commoner's first law of ecology holds: 'everything is connected to everything else' (1971: 29), so saving the part involves knowing what is happening to other parts and to the whole. Co-operation may be possible in a world of bioregions, but reaching agreement may be more difficult under stateless conditions because of the increase in the number of parties to the agreement (Goodin, 1992).

Although economic autarky may be possible, ecosystem independence is not. Trans-bioregional problems render the emphasis on individual ecosystem protection short-sighted. Simply withdrawing from the global economy does not address how to solve existing commons problems (although it may prevent them getting any worse). Bioregionalism may work to maintain global ecosystem health but it is debatable whether it would actually create the conditions required to reach that state of global environmental balance. Given that trans-societal co-ordination and communication are more important within such a context, decisions taken within bioregional communities are only meaningful within that context. Unfortunately, as Taylor admits: 'the controls which can be effective within the small community cannot generally have a great impact on relations between people of different communities' (1982: 167). Added to this is Goodin's observation that 'decentralization gives each member of the community more control over that community's decisions. But the smaller the community, the less and less the community's decisions will ordinarily matter to the ultimate outcome. People are being given more and more power over less and less' (1992: 150). So the multiplication of decision-making units within the context of reaching agreement and co-operation may compromise the self-determination imperative. It is perhaps because of the compromises that trans-communal co-operation necessitates, owing to economic trade and exchange, shared political-normative commitments or ecological commons issues, that bioregionalism is so keen on autarky and linking community tightly to the local ecosystem. Being self-sufficient allows bioregional communities the freedom to determine their own destiny and character.

In this sense Dasmann's conclusion that 'the future belongs to . . . [ecosystem people]' overstates the bioregional case to say the least (in Sessions, 1993: 121). While not wishing to undermine the positive values expressed by the bioregional position, I believe it is not a perspective greens need take *in toto*. There are three main aspects that should be taken from bioregionalism which can be integrated within green politics: firstly, the importance of 'place' and its role as an identity-forming condition; secondly, the emphasis on decentralization and appropriate scale, discussed in Chapters 5 and 6; and finally the extremely useful distinction between 'ecosystem' and 'biosphere' perspectives which is examined in Chapter 6 in more detail.

Social Ecology

Social ecology, although generally sympathetic to, and sharing much with, bioregionalism, offers a different and, in many respects, a more coherent eco-anarchist political vision. Bookchin, the founder and leading theorist of social ecology, calls the social ecology vision of stateless social order 'libertarian municipalism' (1986: 37–44; 1990: 179–85; 1992b). This is defined as 'a confederal society based on the co-ordination of municipalities in a bottom-up system of administration as distinguished from the top-down rule of the nation-state' (1992b: 94–5). It differs from bioregionalism in its concern with the issue of interaction between communities and the rejection of the bioregional model of small-scale, self-sufficient communities (1992a: xix). The confederal nature of the arrangement means it is a voluntary political association of autonomous communities with sovereignty retained at the local level. Yet, the relativism that typified bioregionalism is explicitly ruled out by Bookchin: 'parochialism can . . . be checked not only by the compelling realities of economic interdependence but by the commitment of municipal minorities to defer to the majority wishes of participating communities' (1992b: 97). Here economic–ecological interdependence goes hand in hand with political autonomy and self-determination. Unlike bioregionalism, autarky is not a central principle of social ecology.

Libertarian municipalism as an eco-anarchist theory can be argued to represent a novel form of anarchism. Limiting the scope of communities to simply go their own way marks a decisive break with traditional anarchist thought, which took the communal right to self-governance as its principal and highest political norm. This distinction between libertarian municipalism and other forms of anarchism (including bioregionalism) can be seen in their different understandings of community. Whereas for bioregionalists (Sale, 1980; 1984a; Jones, 1990), and 'pure' anarchists (Taylor, 1982), community is understood as some version of *Gemeinschaft*, libertarian municipalism is presaged on the idea of a 'democratic community'. Community is defined politically not ecologically. Within libertarian municipalism the aim is to recapture the classical political values of the polis and 'authentic' politics, in opposition to the 'inauthentic', modern politics as 'statecraft' (Bookchin, 1992a).

Another important difference is that libertarian municipalism is urban- rather than rural-based (Bookchin, 1992a). Bookchin's understanding of community is thus less 'organic' than traditional anarchist and bioregional views. This marks a significant development in social ecology away from earlier formulations, which accorded normative significance to *Gemeinschaft* and praised the authenticity of organic forms of social life (Whitebook, 1981/2). One could question whether libertarian municipalism is an 'anarchist' theory in the traditional sense of the term. The suggestion here is that it may be better viewed as an attempt to spell out what a more democratized and decentralized society would look like with a continuing role for the state, particularly at the local level.

Bookchin's drift away from pure anarchism is further evidenced by his assertion that there is a 'shared agreement by all [communities] to recognize civil liberties and maintain the ecological integrity of the region' (1992b: 97–8). The contractual and legal-constitutional overtones of confederalism are more usually associated with liberal not anarchist discourse and practice. And his description of the confederal council as composed of elected representatives, with legitimate right to use coercion within a specified ecological territory, to ensure compliance with a shared agreement, could be taken as a traditional Weberian analysis of a state-like political entity, legitimated along standard, but beefed-up, liberal lines (1992b: 99), which is in keeping with the idea that green politics is not 'anti-liberal' so much as 'post-liberal' (Eckersley, 1992a; Doherty, 1996). His assertion that the confederal council, made up of deputies elected in direct democratic elections, is purely administrative with no mandated policy-making powers, which is retained at lower levels (1992a: 297; 1992b: 97), is no more than that, i.e. an assertion. Given the interconnectedness of communities, the existence of a binding confederal agreement relating to human rights and ecological imperatives, and the description of this social arrangement as a 'community of communities', one can imagine the confederal council taking a more proactive role than Bookchin assigns to it. Goodin's criticism of bioregionalism above applies *a fortiori* in this instance, since inter-community relations go beyond trade, or the maintenance of ecological integrity, and consist of substantive normative principles and practices. That is, those who have most power are those who have responsibility for these trans-communal issues.

A weak criticism of Bookchin's position would be that he has failed to clearly and convincingly demonstrate the *stateless* nature of libertarian municipalism. Indeed, by recasting the problem in terms of 'degrees of statehood', rather than in monolithic terms of '*the* state', Bookchin's reformed eco-anarchism is close to themes within recent non-anarchistic radical democratic theory, concerning the importance of plural and decentralized sites of political and social power independent from the state (and the market) (Keane, 1988; Bobbio, 1989). This is particularly evident when Bookchin asserts that, '"the state" can be less pronounced as a constellation of institutions at the municipal level, and more pronounced at the provincial or regional level, and most pronounced at the national level' (1992a: 137), and seems to recommend city and local government levels as appropriate sites for green activism which will not compromise its ends (1992a: 303–4).

The emphasis on appropriate scale is a principle supported by almost all greens. It is usually taken as expressing the need for 'appropriate scale' in political decision-making procedures and especially the sphere of production within which the particular economy–ecology metabolism of the community is located. According to Porritt, 'In terms of restoring power to the community nothing should be done at a higher level than can be done at a lower' (1984: 166). This principle is compatible with state institutions because for some things, particularly international negotiation on global commons issues, the state is the lowest level. The very term 'municipal', with its strongly urban character, resonates and is compatible with the demand to strengthen local and regional tiers of government/governance away from the centre. The principles of libertarian municipalism seem to accord with T.H. Green's assessment of those sceptical of the state. According to Green, 'The outcry against state interference is often raised by men whose real objection is not to state interference but to centralisation, to the constant aggression of the central executive upon local authorities' (1974: 217). Thus the critique of the state is in large part a critique of centralization, and conversely eco-anarchism can be translated as a demand for decentralization and devolved decision-making powers (de Geus, 1996). This demand will be discussed in Chapters 6 and 7 in terms of the local political, economic and ecological levels as often the most appropriate levels for dealing with social–environmental relations.

One can interpret Bookchin's argument for devolving power to municipal levels, yet maintaining a legitimate right for the 'confederal council' to intervene in municipal affairs, as bestowing state-like institutionalized powers on the council. These powers of the council could be regarded as underwritten by 'the shared agreement' (1992b: 98). From this it is not stretching things too far to suggest that this agreement functions as a sort of 'ecological social contract', which on familiar contractarian grounds legitimizes the state. In the manner of a decentralized (and a democratized) state, the confederal council circumscribes communal rights to complete self-legislation, since upholding the ecological compact depends on such circumspection. Political power is *shared* rather than completely *devolved* to local levels, which do not seem possessed of sufficiently strong senses of common *gemeinschaftliche* identity, which could underwrite complete communal self-governance in the manner of pure anarchism or bioregionalism. What I want to suggest then is that the libertarian municipal agenda, the content of which most greens would accept, such as participatory democratic structures, local empowerment, social justice and human rights, is more consistent with a political project aimed at *democratizing the state and civil society* than with abolishing the state. The efficacy of adopting this state/civil-society framework will be defended in Chapters 5 and 6 while the democratic content of this project will be spelt out in Chapter 7.

Eco-anarchism: from Regulative to Constitutive Ideal

Why has the eco-anarchist vision of a federated community of small-scale, face-to-face communities living in harmony with the environment been such an enduring feature of green politics? To answer this question, one must focus on the ideological roots of early green politics.[4]

Firstly, the eco-anarchist vision of the future 'sustainable society' vividly encapsulated, in shorthand form, the basic principles and values of green politics: *inter alia*, ecological and social harmony, decentralization, simple living, quality of life, community and direct democracy. Within the context of the early development of green theory, it was simply assumed that an ecological transformation of society required anarchism updated for the age of ecological limits. Secondly, the dominance of the eco-anarchist vision has to do with the dichotomous style adopted by green theorists and commentators. The most influential instance of this is O'Riordan's fourfold typology of the institutional choices open to green politics: (1) new global order, (2) authoritarian commune, (3) centralized authoritarianism or (4) the anarchist solution (1981: 404–7). In reality the choice comes down to the anarchist solution or the rest, given that it was the only one which, *a priori*, embodied the green values and principles noted above. As green political theory developed, it was assumed that the 'sustainable society' was 'anarchistic' (Dobson, 1990: 83–4).

This utopia-building was prompted, in part, by the need for greens to adopt a strongly critical edge in their analysis of contemporary industrial societies. As Goodwin points out:

> the process of imagining an ideal community, which necessarily rests on the negation of the non-ideal aspects of existing societies, gives utopian theory a certain distance from reality which makes it a sharper critical tool than much orthodox political theory. (1991: 537)

One could say that its initial reaction to the contemporary social world was so antithetical, so radical in questioning almost every aspect, that utopianism was the only form of theorizing which could contain and accurately convey the green message. If asked for proof of the ecological superiority of stateless social order, aboriginal societies and their harmonious social–environmental metabolism could be presented to vindicate the eco-anarchist argument.

This anticipatory-utopian form of political critique is directly related to the evolution of the green movement, and its roots are manifold. Firstly, the practical requirement as a 'new social movement' for it to maintain its distinct identity, and to prevent existing ideologies from stealing its ideas and proposals, presented a good case for accentuating the radical and the utopian. Secondly, in common with other new social movements, greens seemed to be particularly obsessed with questions of self-identity, to demonstrate (to themselves as much as to anyone else) their 'newness' and 'authenticity'. Thus the green movement was at pains to portray itself as a completely new type of politics, 'neither left nor right, but in front'. For example, Porritt declared: 'For an ecologist, the debate between the protagonists of capitalism and communism is about as uplifting as the dialogue between Tweedledum and Tweedledee' (1984: 44), a statement noteworthy for lumping these alternatives together as different versions of the super-ideology of 'industrialism'. Green politics was 'post-' or 'anti-industrial', which cast greens as the vanguard of the future society (Milbrath, 1984), and green politics as the politics of the twenty-first century (Sessions, 1995a).

Thirdly, added to these internal dynamics was the simple fact that, as a new social actor, it had little or no access to the policy-making process, and therefore did not need to outline programmes, budgets or detailed policies in the language of that process. Broad-brush strokes rather than attention to the fine print characterized early green discourse. The over-riding imperative was to distance itself theoretically from the reality surrounding it, and in the case of the 'eco-monastic' strand of eco-anarchism, to turn one's back on the existing social order and create 'liberated zones' from the 'industrial mega-machine' (Eckersley, 1992a: 163–7; Bahro, 1994). This formative experience, like all formative experiences, still exerts an influence on green politics.

Fourthly, this concern with outlining its diametric opposition to the status quo was underpinned by the rather naive belief that the green case was so obvious and so compelling that all that was needed was simply publicly to outline its critique and proposed solutions (Dobson, 1990: 131). Finally, following the 'doom and gloom' that typified the post-*Limits to Growth* (Meadows et al., 1972) ecology movement, there was clearly a need for greens to outline an image of a better future. As Paehlke points out, 'The Malthusian perspective is neither necessary nor helpful in engendering positive change. An environmental perspective and policies must seek to create a *preferable* world' (1989: 55, emphasis in original). Like a skilled preacher the early green movement had threatened apocalypse if its warnings were not heeded; now it also promised the eco-anarchist, liberated society if people changed their ways in time.

A cursory review of green literature will quickly highlight the extent to which green ideologists and commentators are obsessed with presenting the green case in an either/or simple binary format. Almost ubiquitous is the habit of drawing lists distinguishing 'green' from 'non-green' positions. Examples include Porritt's two 29-item lists differentiating 'The politics of industrialism' from 'The politics of ecology' (1984: 216–17), O'Riordan's (1981) technocentric/ecocentric dichotomy, Dobson's distinction between 'ecologism' and 'environmentalism' (1990: 13), and Capra's 'paradigm shift' from 'The Newton World-Machine' to

'The New Physics' (1983: Part II). Surprise that this dualistic methodology is so widespread within a political theory that is supposed to be holistic is only surpassed by the fact that it persists to frame its concerns. This dualistic thinking (also noted in the discussions of deep ecology and non-anthropocentrism) is clearly evident in the debate about the role of the state within green politics. From the eco-anarchist perspective there is a simple (and simplistic) choice: *either* centralization *or* decentralization, anarchism *or* the nation-state, ecology *or* industrialism (Carter, 1993). That green political theory could attempt to combine elements of both is pre-emptively dismissed as reformist and therefore not 'really' green.

It is from this dualistic methodology, coupled with the utopian-critical demands of the early green movement, that eco-anarchism became the dominant political theory of greens. Three steps can be identified in this process:

1 the concern with what a 'sustainable society' would look like, to highlight the unsustainable nature of existing society; which led to
2 a focus on mapping 'the sustainable society', that is describing, often in great detail, a generally agreed picture/blueprint of that society;
3 the assumption of the sustainable society as 'anarchistic', to rule out eco-authoritarian dystopias, and to act as the benchmark against which 'greenness' could be judged.

It is the particular historical development of green theory (both internal debates, and between it and other theories such as socialism and liberalism), and green political practice (fundis/realos), that largely account for the prevalence of the eco-anarchist solution. These factors produce a marked tendency within green theory to work backwards, as it were, from utopia to theory, with practical engagement in the political realities surrounding it reduced to publicly articulating the utopian-theoretical synthesis. Although there is nothing wrong with outlining a vision of a better society – indeed this prescriptive dimension is the mark of any ideology worth its salt – this tendency unfortunately resulted in the description of the future society becoming a substitute for the task of specifying and spelling out green principles. These principles were held to be self-evident and could be 'read off' from the description of the future eco-anarchistic society. If however we start from green principles and values there is no reason to believe that a society consistent with them (assuming that these values are compatible with one another) will necessarily be 'anarchistic'. It is perhaps more than coincidence that the common 'reading off' of social principles from nature often occurs together with 'reading' them off from the anarchistic society: the one reinforces the other (Sale, 1980: 329–35; Dobson, 1990: 24–5; Bookchin, 1991: 75–86).

Writing in 1990 Dobson suggested that 'the Green sustainable society can be negatively defined by saying that it will not be reached by transnational global co-operation, it will not be principally organized through the institutions of the nation-state, and it is not authoritarian' (1990: 84).[5] However, if one starts from green principles, it may be that some of these institutions, values and practices are required for the realization of green political goals. It is arguably for the latter reasons that by 1995, in the second edition of *Green Political Thought*, Dobson concluded that 'the possible political arrangements in a sustainable society seem to range all the way from radical decentralization to a world government' (1995: 123). As many are beginning to realize, determining what a green society would look like (green ideology) is a poor substitute for articulating and justifying green political principles (green political theory) and proposing green policies.

Conclusion

What this chapter has argued is that eco-anarchism is not an essential component of green politics, in that the values greens espouse may be institutionalized in non-anarchistic ways. For example, the eco-anarchist concern with autonomy and self-determination is something which as a green value can be realized in non-anarchist ways. Autonomy is discussed in Chapter 6 where an ecological virtue perspective on human flourishing is argued to hinge on the relationship between human autonomy and welfare. Nor is it desirable that, as it stands, the eco-anarchist utopia acts as a fetter on the future development of green theory, unnecessarily precluding its positive engagement with the state. It is perhaps not completely contingent that a reassessment of eco-anarchism within green theory is occurring at a time when the minds of greens are turning from ideals to principles, and from principles to practice. This is not to say that eco-anarchism is to be ejected from the green political canon: the integration of its insights within the context of green theory moving from negative criticism to positive proposals calls for it to become a regulative rather than a constitutive ideal for green politics – that is, informing and guiding, but not determining its goals. Thus while eco-anarchism may not have all that much to offer by way of green thinking about possible institutional structures for a sustainable society, it does have much to offer by way of what one can call a 'cultural' (including inspirational) contribution to green political theory, particularly, as Eckersley (1992a: 186) points out, when eco-anarchism shifts from a utopianism of form to a utopianism of process.

Institutional arrangements are thus to be judged instrumentally in terms of whether they hinder or promote green practices and values, the sum of which I term 'collective ecological management'. On this reading it is the 'essentialist' view of the state held by eco-anarchists like Sale (1980) and Bookchin (1991), who regard the state as *intrinsically* inimical to the ecological and other social values espoused by greens, that grounds their rejection of the state as a part of the green political project. A good example of this is Khor's argument that 'under state control the environment necessarily suffers' (in Goldsmith et al., 1992: 128). This is especially clear in Bookchin's thought. The originating thesis of social ecology is that the ecological crisis is due to hierarchy. The domination of humans over nature is the first level of this hierarchy but Bookchin (1991: 2–12) argues that this hierarchical relationship itself stems from the domination of humans by other humans. For him, as the state is the highest contemporary expression of social hierarchy, it is the ultimate cause of the present ecological crisis. Added to this is his view that the 'state' is not just a set of institutional arrangements but also a psychological disposition. According to Bookchin, 'the State is not merely a constellation of bureaucratic and coercive institutions. It is also a state of mind, an instilled mentality for ordering reality' (1991: 94).[6] For Bookchin, as for others in the anarchist political tradition, this 'instilled mentality' is a combination of unreflective subservience, apathy and powerlessness. These are extremely strong claims, to say the least, the plausibility of which really depends upon accepting the anarchist analysis as a whole, particularly its version of the historical origins and evolution of the state (Carter, 1993). It is because of their essentialist conception of the state that eco-anarchists such as Bookchin (1991; 1992a) argue that the resolution of the ecological crisis is simply *impossible* while the nation-state exists. But more than that, on traditional anarchist grounds the state is also deemed to be *unnecessary* as well as *undesirable* to its resolution. The plausibility or otherwise of the anarchist position need not detain us, since as argued previously, the eco-anarchist perspective is to be thought of as a *constitutive* as opposed to a *regulative* ideal of green political theory. It is this 'essentialist' view of the state that explains its rejection on eco-anarchist grounds. If this essentialist view is rejected, then the eco-anarchist solution does not constitute an insuperable barrier to a positive green engagement with the state.[7]

The conclusion of this chapter, that an immanent critique of the state rather than its rejection is more appropriate to green political theory, is similar to the immanent critique of the Enlightenment and anthropocentrism suggested in previous chapters. The problem of eco-anarchism's 'utopian' critique is that it is a 'view from nowhere'. That is to say, the values and principles it represents are not widespread within the existing culture. As Hayward puts it, 'critique [becomes] mere criticism [when it] appeals to a utopian vision that others may not share, which is not rooted in the norms and values of the culture, and so is an abstract "ought"' (1995: 51). The point about immanent critique is that it starts from where we are now, rather than adopting a view from nowhere, a view from the past or a view from the future. That is, we can only approach the 'new' via a critique of the old, rather than simply think up wonderful blueprints for the future. Immanent critique represents a qualitatively different kind of theorizing from utopian critique. While it is less 'radical' in the sense that it is committed to the possible and not just the desirable, it is all the more radical in the sense of being a realizable alternative to the status quo.

One may view eco-anarchism as a permanent reminder of the dangers and problems involved in the state having a role in social affairs and ecological management. At the same time eco-anarchism may also be understood as emphasizing that the state's role in ecological governance is a *necessary* rather than a *sufficient* condition for achieving green goals, i.e. reflecting the state's instrumental as opposed to intrinsic role and value (Barry, 1995a). However, the argument for the state having a key role in providing environmental public goods has not been undermined. It is to an examination of the state that we turn next.

Notes

---

1 The eco-anarchist goal of transparent social relations is similar to the deep ecology goal of transparent moral relations with nature (that is the priority accorded to direct appreciation of nature and its search for the 'truth' of human–nature relations). This deep ecological attempt at transparency is also a vice of Enlightenment science and technology, based on the Baconian hubris that human reason can fully comprehend that which it has not created. As suggested in the last chapter, there are limits to human knowledge of the world. The issue of social transparency in relation to representative versus direct democracy is discussed in Chapter 7.

2 According to Taylor, 'the liberal theory of the state critically depends on the assumption that individuals are pure egoists or at least are "insufficiently" altruistic; with enough altruism, this rationale for the state evaporates' (1982: 55–6). As I argue in Chapter 7, a weaker form of this can be defended in terms of environmental considerations leading to a view of democracy in which individuals take the interests of others into account. However, while this can be seen as an attempt to encourage people to be less 'unecological', this is viewed not in terms of abolishing the state but rather in terms of a virtuous green citizenry to complement the democratization of the state along green lines.

3 According to Georgescu-Roegen, 'the absence of any difficulty in securing raw materials by those countries where modern economics flourished was yet another reason for economists to remain blind to this crucial economic factor [nature's perennial contribution]' (1971: 2). This logic of displacement, whereby the necessities for economic development are provided not by one's own local ecosystem, but by non-local ecosystems, continues in the modern age with the advent of the global market where the planet's resources are traded as commodities, and thus in theory available to any society. In the contemporary world, 'effective demand' (purchasing power) rather than 'local supply' (nearness of resources) determines the ecological inputs available to any particular economic system. Thus the link between the local economy and the local ecosystem is broken, and in the process (as critics of 'globalization' have argued) the distinctiveness of local cultures is undermined.

4 The distinction between constitutive and regulative ideals is taken from Kant, 'A principle is "regulative" when it merely guides our thinking by indicating the goal towards which investigation should be directed . . . it is "constitutive" when it makes definite assertions regarding the existence and nature of the objectively real' (1957: 211).

5 A problem with this formulation is the implicit assumption that the issue of political power and social order within a nation-state framework necessarily calls for 'authoritarianism'. That authority can be legitimate and democratically accountable also needs to be considered before condemning the state as 'essentially' authoritarian and exploitative.

6 Despite their many and profound disagreements, and Bookchin's often vicious criticisms of deep ecology, social ecology and deep ecology do share some central positions. Firstly, both accept the centrality of psychological transformation to the green project: deep ecology emphasizes the shift from an anthropocentric conception of self to a wider ecological sense of self (see Chapter 2), while social ecology stresses the substitution of an unreflective obedience to state authority with a rational belief in one's capacity for self-determination and autonomy. Secondly, following a tradition begun by Rousseau, both base their respective critiques of the status quo on the *artificiality* of contemporary institutions and ways of thought and life. Deep ecology (and bioregionalism) claims that modern social life is 'divorced' from its 'natural context', and out of touch with the rhythms of the natural world. In a similar vein, social ecology makes much of the 'artificiality' of the nation-state in opposition to the 'natural' forms of human sociality expressed in stateless forms of human organization. Hence Bookchin's (1992a) distinction between the authenticity and naturalness of 'politics' and the artificiality of 'statecraft'. Thirdly, both work with an explanatory ideological framework within which the idea of 'the fall' and the possibility of 'redemption' are central orientating concepts. For deep ecology the Enlightenment, modernity and industrialization constitute humanity's 'second fall' while for social ecology 'the State' is society's 'original sin' (Bookchin, 1991: 2).

7 However, even if the anarchist argument is on empirical or historical grounds, the fact that the state may behave in 'oppressive' ways does not 'prove' the state's 'essentially oppressive' character, or that it cannot be used for non-oppressive purposes.

5

The State, Governance and the Politics of Collective Ecological Management

CONTENTS

The previous chapter discussed and raised some doubts concerning the centrality and role of eco-anarchism within green political theory. This chapter seeks to outline an alternative institutional programme for green politics. A central aspect of this alternative view is the positive role given to the nation-state in the resolution of the various ecological, economic and ethical dilemmas raised by green politics. In what follows I propose an instrumental view of the state within the context of what I term 'collective ecological management'. That is, the state should be seen not as a *green value itself* but rather as one particular institutional *means* through which green values and practices can be realized politically.

In Chapter 3, I argued that the 'ecological niche' for humans is created rather than naturally given and that a 'humanized' or transformed environment is our 'natural' habitat. The collective management, manipulation and intentional transformation of the environment are thus universal features of all human societies. As a universal requirement they are, in a sense, pre-political. It is *how* human societies create their humanized ecological niches, the various institutional mechanisms used to maintain a stable metabolism between the social and the natural system, that raise moot political and moral questions. In this chapter, 'collective ecological management' is presented as an institutional form regulating this metabolism based on green values and principles. This idea of active ecological management cuts across the deep–shallow, radical–reformist continuum within green theory. What conceptions of green political theory differ over are the scale, type, institutional structure and normative side-constraints operative upon social–environmental metabolic states, not the necessity for environmental management and transformation. For example, even deep ecologists, for whom a pre-emptive 'hands-off-cum-nature-knows-best' position constitutes a central principle, accept that preserving wilderness requires active social, and particularly institutional, intervention. In other words, preservation from development, as much as conservation for (future) development or ecological restoration, all take place within the broad framework of 'ecological management'. The deep ecology ideal of wilderness preservation, the preservation of the non-human world from a certain type of collective human transformation (in the form of 'development'), paradoxically necessitates another form of human management, in the form of institutional structures, practices etc. which function as a form of social governance to limit and/or transform development, such that wilderness is preserved. What appears as non-management at one level is at another level simply another form of management. 'Walking lighter on the earth' is as much a form of ecological management as economic development. The political and normative issue is that collective purposive-transformative interaction with the environment can simply be more or less extensive, have a different character or be more or less sustainable.

In Chapter 3, the normative issue relating to social–environmental interaction was argued to concern the boundary between 'use' and 'abuse'. This chapter builds on that position, a position which sees the normative core of green politics as concerned with an 'ethics of use'. The political issue for green politics is over the type, scale and institutional structures for managing social–environmental interaction which best concur with green values and principles. What this chapter sets out to explore is the role the state may play in this process; it does *not* set out to provide a green justification for the state. Rather the institutional focus of this chapter should be taken as an attempt to widen the parameters of the debate around the political, economic and cultural structures of a sustainable society beyond that offered by a stark choice between the status quo or eco-anarchism. More precisely, the aim of this chapter is to further develop this understanding of green political theory by discussing the state as a particular institutional form, and its role and limitations in realizing green goals. These goals can be summarized as ecological sustainability and morally regulated collective interaction with nature.

The institutionalization of green values within collective ecological management should be thought of in terms of 'governance' as opposed to 'government'. That is, while the state has a role to play in creating and maintaining a sustainable and ethically informed metabolism with the environment, it is not the only or the pre-eminent institution in this process. Although aspects of a theory of a 'green state' may be gleaned from this chapter, the instrumental view of the state from which governance starts, which sees the state as one amongst other institutional forms that regulate a specific social–environmental metabolism, ought to temper an interpretation of the position being defended here as overly 'statist' (Wall, 1994; Paterson, 1995). Such an interpretation reveals the dualistic thinking that characterizes the eco-anarchist standpoint, where the choices for green politics are either 'statism' or 'eco-anarchism', the state or community, and so on. The principle of *appropriateness* with regard to institutional design, scope, scale, co-ordination and internal regulation is the touchstone of collective ecological management. Institutional arrangements ought to be judged by their appropriateness for achieving green values and practices. There is nothing new here since 'appropriateness' is a well-established green institutional principle (Porritt, 1984: 164–5; Dobson, 1990: 125; Martell, 1994: 54–8). As in the search for a conception of green moral theory which transcends the non-anthropocentrism/anthropocentrism dualism, this chapter seeks to widen out the issue of institutional arrangements within green politics beyond the opposition between 'eco-anarchism' and 'eco-statism'. A good example of this is Wall's claim:

> The debate . . . over the governance of Green societies has been a debate between eco-anarchists and eco-statists, while one party claims that the creation and maintenance of Green imperatives demands centralised restraint, the other argues that such imperatives are served by greater freedom, participation and self-government. (1994: 13)

In this chapter, like Wall, I will discuss the question of institutional design and organization within green politics in terms of 'governance'. However, unlike him, I discuss institutions here in the light of their efficacy in realizing green values and principles, such as sustainability, rather than how they fit into some ideal of the 'sustainable society'. At the same time, the supposition that the only means available to state governance is 'centralized restraint' (with the implicit suggestion that non-state forms of 'decentralized restraint' are either impossible or not really forms of restraint at all)[1] will be challenged, as well as the simplistic binary presentation of the institutional options available to green political theory.

In some respects the instrumental view of institutions, including the state, adopted in this chapter is close to Goodin's (1992) division of green political theory into a 'green theory of value' and a 'green theory of agency'. According to him, the green theory of value 'provides the unified moral vision running through all the central substantive planks in the green political programme' while the green theory of agency 'advises on how to go about pursuing those values' (1992: 15). In Goodin's view, these two aspects of green politics are logically independent, such that accepting green arguments concerning the intrinsic value of the natural world, for example, does not imply accepting green arguments concerning political arrangements. As he puts it, 'we can, and probably should, accept green policy prescriptions without necessarily adopting green ideas about how to reform political structures and processes' (1992: 5). While rejecting the thrust of Goodin's particular dualistic analysis of green political theory, I do wish to draw a similar means–ends distinction for purposes of exposition. The heuristic distinction I wish to make is between green values and aims (such as sustainability, democratization and the moralization of social–environmental interaction) and the various institutional arrangements and social practices that may frame, articulate, embody, prioritize or otherwise realize these values in determinate social and social–environmental relations. On the one hand, green values are perhaps compatible with a narrower range of institutional arrangements than Goodin allows, such that accepting 'green values' does commit one to adopting some distinctively 'green' political and economic structures. On the other hand, in opposition to so-called 'radical greens', it is also possible that green values are compatible with a wider range of institutional arrangements than they acknowledge, including the state. This chapter looks at the institutional options available to achieve green goals in the space between these two positions. Again in opposition to Goodin, this chapter, by building on the ethical arguments raised in Chapters 2 and 3, seeks to present a conception of green politics in which its ethical and political commitments are both directly rather than contingently related *and* mutually supporting.

Structure and Agency within Green Politics

From the point of view of green political theory, the resolution of ecological problems is not simply a matter of structural reorganization, either of the economy and the scale of technology and production, or of changing the level of legislative or policy-making power. For many of the issues that green political theory deals with, particularly those related to its moral claims, attention needs to be focused on changing people's attitudes, interests and modes of acting. To put it simply, many of the questions raised by greens are matters of individual and collective will as much as institutional transformation. They are of course related, and the specific manner in which agents are related to structures will constitute a particular understanding of green theory. For example, we can understand deep ecology as a theory which emphasizes agent-level change (consciousness raising) which works on the assumption that the appropriate structures will 'follow' from this deeper-level change. The argument here is that, for green politics, institutional change (politically, economically, socially) must be placed within a wider cultural context.

As we have seen in previous chapters, a non-anthropocentric green politics aims to alter the prevailing *attitudes* to nature. On this account, green politics is a moral crusade seeking to win 'converts' away from anthropocentrism and its worldview. These approaches have much in common with what Dobson calls the 'religious approach' to green change which holds that 'the changes that need to take place are *too profound to be dealt with in the political arena*, and that the proper territory for action is the psyche rather than the parliamentary chamber' (1990: 143, emphasis added). On the other hand, so-called 'reformist environmentalism' focuses mainly on 'greening' existing structures, rather than reflecting on the structures themselves in the light of ecological considerations, and the relationship between structures and the behaviour and attitudes of agents.

The conception of green political theory being developed here seeks to combine both agency and structural approaches. Indeed, as I hope to show in this and the following chapters, the values, principles and goals that are central to the green political project depend upon combining both agent- and structural-level change. Collective ecological management therefore has to do with both preferences and policies, agents and structures. In this respect, unlike market approaches to ecological issues, collective ecological management is a problem-solving rather than a preference-aggregating process. For green politics, understood as a form of collective ecological management, resolving environmental problems requires *cultural* and not just institutional change. Because the roots of ecological problems do not lie exclusively in either cultural norms or institutional structures, neither do the solutions. From the point of view of green politics defended here, the long-run resolution of social–environmental problems requires a politics based on an *immanent critique* of the prevailing cultural as well as institutional order.

From a green point of view, social practices form an important connection between structures and agents, as well as between agents themselves. MacIntyre describes a practice as:

> any coherent and complex form of socially established co-operative human activity through which goods internal to that form of activity are realized in the course of trying to achieve those standards of excellence which are appropriate to and partially definitive of that form of activity. (1984: 187)

In terms of the analysis of virtue earlier, it is clear that MacIntyre's 'standards of excellence' are another way of spelling out virtues, although I do not subscribe to his particular interpretation of virtue. At the same time, I want to suggest that a green examination of the institutional structure appropriate to its values requires attention to the way in which practices are related to institutions. As O'Neill points out, 'Institutions, however, not only sustain practices, they can also corrupt them. The pursuit of external goods – wealth, power and status – may come into conflict with the pursuit of internal goods and practices' (1993: 127). Social practices are important for green institutional arrangements. However, it is not the case that green institutional arrangements are to be determined by social practices. Rather it is that institutions are to be judged to the extent that they support rather than undermine practices which embody green values, principles and modes of behaviour. At the same time it is not only practices, in the strong sense implied by MacIntyre, which can express values or be partly constituted by normative considerations. Institutional change can also signal profound changes in modes of behaviour. As Dryzek notes, 'in remaking our institutions we also remake ourselves: who we are, what we value, how we interact, and what we can accomplish' (1987: 247). While it is the case that green institutional arguments favour structural changes which enhance ecological social–environmental practices (which focus on 'internal goods'), this does not rule out social–environmental exchanges which are governed by 'external goods' (such as wealth creation). Both internal and external goods can contribute to human well-being. It is not the case that all social–environmental practices are automatically exempt from criticism, and all institutionally regulated social–environmental exchanges ethically wrong. That is, not all social practices are assumed to be morally symbiotic, nor are all institutional activities assumed to be morally parasitic. Rather, an ethics of use, as a publicly agreed set of normative codes which distinguishes 'use' from 'abuse', should be thought of as ranging over all social–environmental exchanges, whether they be institutional (i.e. mediated via the state or market), individual or carried out within specific ongoing social practices.

Thus, even if we allow for the importance of 'ecological consciousness' as an essential aspect of green politics, as deep ecologists and eco-anarchists claim, the state does not necessarily stand in the way of the spread of such mores and modes of being.[2] The point is that it is through a self-reflexive awareness of the relationships between individuals as citizens, consumers, producers, parents, in reference to an understanding of themselves as ecological beings within the spheres of the nation-state, civil society and the economy, in reference to the overall ecological context of these spheres, that ecological modes of action and thought can be realized and ecological virtue cultivated. It is in social practices and social institutions that green norms and ecological modes of behaviour will be located. The point about collective ecological management is not simply to do with finding a more effective institutional response to environmental problems. Rather, it seeks to tackle the causes directly and not just the effects of social–environmental problems, by expanding the criterion of 'effective' to include normative as well as 'instrumental', utilitarian or narrowly economic considerations. Crucially, it has to do with meshing the activity of structures and agents in such a way that the collective outcome is ecologically rational and socially acceptable.[3] In other words, structures and agents are seen as dialectically (and democratically) related, such that agents (collectively) can have some input at the level of structures which themselves influence the behaviour of agents by affecting the conditions under which they make decisions.

Green politics in the last analysis is not simply about macro-level changes, but is also about choosing to live in a different manner at the micro-level of individuals and communities. Building on the discussion of virtue earlier, within collective ecological management, individuals are faced not just with the question of 'what ought I to do?', but more importantly have to ask themselves 'what sort of person do I wish to be?', and ultimately 'what sort of society do I wish to be in?' In response to Hume's adage that we ought to 'design institutions for knaves', one desired aim of green politics is to encourage people to be less knavish in the first place. This requires designing institutional structures to sustain 'ecologically rational' modes of behaviour by supporting rather than undermining ecological social practices. It is my contention that ecologically rational modes of interaction involve, in part, cultivating ecological virtue at the individual level in the various roles people play and identities they have as consumers, producers, citizens and parents. At the social level, ecologically rational social–environmental relations ultimately require the creation of an ecologically adapted and adaptive culture, supported by an institutional structure in which the state and market are restructured so that they are instrumental to social life and democracy as popular sovereignty. It is to this criterion of ecological rationality that we turn next.

Ecological Rationality

In order to assess what institutional forms best concur with the conception of green politics being developed here, a criterion or standard by which they can be judged is required. To this end I adapt Dryzek's criterion of 'ecological rationality' to my own purposes in order to develop a workable, heuristic device by which to rank alternative institutional arrangements for dealing with social–environmental relations. According to Dryzek ecological rationality can be understood as:

> the capability of ecosystems consistently and effectively to provide the good of *human* life support . . . From the perspective of ecological rationality . . . what one is interested in is the capacity of human systems and natural systems *in combination* to cope with human-induced problems. (1987: 36, emphasis in original)

This idea of ecological rationality, which captures the essence of ecological sustainability, will be used as an appropriate criterion against which to judge different institutional arrangements. However, ecological rationality, as Dryzek defines it, is problematic from the green point of view defended here, since it refers only to 'human life support' with no reference to other values such as democracy, autonomy or social justice. Nor does it refer to the interests of the non-human world. Dryzek's view of ecological rationality may be said to be a 'pure ecological sustainability' conception, concerned simply with maintaining a stable and productive entropic metabolism between social and ecological systems. That is, ecological rationality for Dryzek is a form of functional rationality (1987: 34–5). This is not surprising since he discusses social–environmental interaction in terms of the 'human system' interacting with the 'natural system', with the aim of securing long-run human 'life support'. However, within collective ecological management the appropriate criterion is a conception of ecological rationality which includes Dryzek's 'sustainability' conception but expands it to include the normative as well as the 'functional' dimensions of social–environmental problems. That is, this expanded view of ecological rationality refers to moral/political (i.e. parasitic and symbiotic, use and abuse, just and unjust) as well as functional (sustainable and unsustainable, productive and unproductive) criteria by which to judge social–environmental relations. This can be seen to be a necessary consequence of viewing green politics as concerned with *social–environmental relations* rather than simply the *economy–ecology metabolism*, which is a subset of those relations.

Thus this expanded understanding of ecological rationality is one which attempts to address the totality of social–environmental relations, not just the material aspect of our interaction with the non-human world. The expanded conception refers to the impact of human and natural systems upon non-human welfare as well as being concerned with values other than long-term life support. As used here, ecological rationality also refers to *both ends and means*. That is, in sum, an ecologically rational social–environmental metabolism must fulfil three interrelated criteria. Firstly, this metabolism needs to be sustainable, in the sense indicated by Dryzek. Secondly, it must be morally symbiotic, in the sense indicated earlier. And finally, it must be socially acceptable, i.e. decided and maintained democratically rather than undemocratically, extended to a point where the ends as well as the means of social–environmental exchanges are on the political agenda. The range of issues covered by the expanded conception of ecological rationality is therefore greater than that covered by a pure sustainability conception. Whereas the instrumental view judges social arrangements in terms of their ability, in conjunction with ecosystems, to produce long-term sustainability of a given view of human welfare and life support, the substantive view judges arrangements with respect to values other than human welfare.

The resolution of environmental problems from a green point of view involves normative as well as practical considerations. Environmental problems, even when presented as economic problems of scarcity or underpricing, or political problems of legitimacy, competence or whatever, are at root deeply normative as well as technical problems. That is, these problems are not just about the social–environmental means which sustain human welfare, but are also about what human welfare means, and whether considerations of human welfare alone ought to regulate social–environmental relations. Green politics is concerned with initiating a public debate over the ends of the social–environmental metabolism, rather than debate over the most effective means to given ends. It is the green contention that much of the environmental crisis stems from the *depoliticization* of the ends of social–environmental interaction, the prime example of which is the central and agenda-setting place accorded to undifferentiated economic growth and consumption within Western societies, politically, economically and culturally (Barry, 1990; Beck, 1992).

The point is that from the green point of view the ecological crisis is not just a technical problem which requires the adoption of a specific set of institutional arrangements or social choice mechanisms (although these are of course crucial). From the point of view of green political theory, the 'ecological crisis' is not only a crisis *for* society but also a crisis *of* society in the sense of stemming from contradictions *within* society. As a normative problem the appropriate criterion against which any 'solution' should be judged calls for substantive and not instrumental rationality. It is for this reason that economic solutions to environmental problems are questioned.[4] For example, both economic rationality and ecological rationality *qua* sustainability are governed by the principle of efficiency geared towards maximization, without any normative consideration of ends or the impact on non-human welfare. Ecological rationality as a communicative form of rationality in respect to the economy–ecology metabolism is not geared towards the long-term maximization of human material welfare alone. Societies do manage their environments so that ecosystems are constrained to an 'anthropogenic sub-climax' (Odum, 1983: 473).[5] According to Dryzek, '*in the absence of human interests*, ecological rationality may be recognized in terms of an ecosystem's provision of life support to itself. Left to its own devices, ecological succession tends toward the production of climax ecosystems' (1987: 44, emphasis in original). Thus within sustainable social–environmental relations, ecosystems are characterized by stable sub-climatic states and unsustainable relations by unstable sub-climatic states. However, symbiotic-sustainable relations will be characterized by sub-climatic ecological states within which long-run sustainability (in terms of human material welfare) will be less than that given by a pure sustainability conception of ecological rationality. In other words, whereas sustainability implies human society constraining and managing the natural progression of ecosystems for its own interests in maximizing long-run human material welfare, symbiosis implies society imposing *extra*, normatively based, limits on itself in the sustainable use of ecosystems. For collective ecological management, functional rationality is not sufficient by itself (though it is necessary) for resolving ecological problems. From the point of view of green politics, resolving the ecological problems facing society requires deliberative processes within which the normative content and ends of social–environment relations may be discussed, debated and hopefully reconstituted. However, this is not to say that the expanded conception of ecological rationality outlined here will *guarantee* sustainable and symbiotic social–environmental relations. Nothing can guarantee the latter and seeking to frame green politics in those terms is counter-productive. All one can reasonably hope to achieve is to create the context within which it is *more likely* that the social–environmental metabolism will be ecologically rational. Hence the importance of extra-institutional, cultural change for green politics.

A final way in which to understand ecological rationality is to see it as an indication of the learning and adaptive capacities of social institutions to cope with the material and moral dimensions of the social–environmental metabolism. Ecological rationality, referring as it does to the normative as well as material aspects of social–environmental interaction, is in part concerned with the interrelationships between the various parts that together constitute a particular pattern of social interaction with the non-human world. As such, collective ecological management (which has ecological rationality as its principal criterion) is about the ways in which economic, political, normative and cultural valuations of the environment, the particular social–environmental relationships they indicate, and the human interests evoked, relate to one another. In the next section, the state as a central institution of collective ecological management is examined.

The State and Ecological Management within Green Politics

Some green theorists and commentators have sought to argue that the election of a 'green government' be seen as an interim measure, an intermediate stage on the way to the future sustainable society (Young, 1994; Spretnak and Capra, 1985). This commitment to state institutions, and the political aim of 'greening the nation-state', are often found in debates about strategy within the green movement. For example, Irvine and Ponton explicitly state that the aim of their book, tellingly entitled *A Green Manifesto*, is to 'explore how governments might begin to move towards this [green] goal, and away from the present slide into ecological and social chaos' (1988: 16), and go on to offer detailed policy prescriptions for an incoming 'Green government' (1988: 30). Other examples of this strategic aim to greening the state can be found in Porritt (1984: 165–7) and the programme of the German Green Party (Die Grünen, 1983). Sometimes the use of 'statist' means to achieve non-statist, long-term ends is argued to be compatible with eco-anarchism. A case was made for such an interpretation of Bookchin's libertarian municipalism in the last chapter. A similar view may be found in Begg's argument that 'Complete rejection [of the state system] would clearly be a mistake, little of this [green political change] is possible without the incapacitation and *internal transformation* of the centralized state' (1991: 29, emphasis added). This instrumental-strategic view of the state as a stepping stone to, and a possible institutional part of, the sustainable society is one that will be examined in this section.

There are also more directly policy-orientated conceptions of green politics which work with the assumption of state institutions as given, often in an uncritical fashion. The policy orientation almost by definition means these approaches have to frame their proposals with the assumption of the continuing importance of the nation-state, since state institutions are the primary policy-making and implementing agencies (Kemball-Cook et al., 1991; Weale, 1992; Young, 1992; 1993). A recent example of this is 'ecological modernization' which is discussed in the next section. The centrality of the state is perhaps most obvious in policy-orientated debates on the international dimensions of environmental problems where the nation-state is simply accepted as the appropriate institutional subject of analysis (Goodin, 1992: 146–68). Together with the strategic acceptance of the state outlined above, the tensions caused by the pragmatism of policy-orientated green political action are readily seen in the division between 'realos' and 'fundis' within the green movement (Doherty, 1992). This division is partly a debate concerning different means to shared ends. Realos defend the 'greening of the state' as an intermediate step to the end of the sustainable society, while fundis claim that one cannot use statist means to non-statist ends. A variation of this argument concerns those who place the role of the state in environmental affairs within the context of the multi-institutional 'governance' that reaching and maintaining a sustainable development path for society requires (Jacobs, 1995). The latter is close to the collective ecological management position being developed in this chapter.

A third reason for 'green statism' can be found in the writings of those for whom the green endorsement of state regulation is a mark of their scepticism, if not outright rejection, of market-based approaches to environmental problems. These range from the 'green social democracy' of Eckersley (1992a; 1992b) and de Geus (1996) to the eco-socialist proposals of Stretton (1976), Mulberg (1992; 1993), Weston (1986) and Ryle (1988), and the eco-Marxism of Harvey (1996) and Pepper (1993). Such arguments for the state can be traced back to the initial political debate on the environmental crisis in the 1970s. Typical of this early green suspicion of market-based approaches is Ashby's statement that 'The future of man's [*sic*] environment cannot be left to private enterprise . . . Therefore governments have to take responsibility' (1974: 6). Here much of the argument focuses on either the limited ecological rationality of the market (Dryzek, 1987), or the inability of the market to deal with the question of the optimum scale of the economy relative to the environment (Daly and Cobb, 1990), or its inability to cope with the degree and type of structural change that sustainability requires (Jacobs, 1994; 1995). A related critique developed in the next chapter concerns the unsuitability of market approaches to articulate the normative and political dimensions posed by ecological problems.

A related argument in favour of the state turns on the green recognition of the utility of state institutions with regard to administering social justice, itself based on the intrinsic connection between green aims, such as ecological sustainability, and social justice. An early example of this is Daly's (1973a) argument that a steady-state economy, a necessary consequence of the early 'limits to growth' argument, which is characterized by a fixed amount of wealth, increases the necessity and desirability for state redistribution. More contemporary arguments in favour of the state in terms of distributive justice have to do with the incapacity of market and/or anarchistic institutional arrangements to ensure an equitable distributive pattern from a green point of view (Gorz, 1982; Eckersley, 1992a: 175–8; Goodin, 1992: 150).[6]

A final argument in favour of there being some role for the state within green politics comes from those writers concerned about the status of democratic norms and practices within green political theory. Typical of this position is Frankel who argues, from an eco-socialist position, that 'any post-industrial society is going to fall far short of achieving greater equity, social justice and genuine popular sovereignty, without the construction of a democratic state structure' (1987: 51–2). A similar position is held by theorists such as Saward (1993) and Wissenburg (1998), for whom representative democracy and the liberal democratic state can achieve green goals. In the latter case, the retention of the state for achieving sustainability is, in part, a defence of liberal democracy.

Some proposals for a 'green state' are naive in the sense that they see the institution of the state as a neutral mechanism which can be used unproblematically to forward the green policy, and in some cases, normative agendas. For the most part, however, advocates of the 'green state' are pragmatists who accept that the state is an institution without which green proposals cannot be realized. That is, most of those who emphasize the role of the state within 'ecological governance' see it as one of the primary institutions with the capacity and legitimacy to carry out, either directly or indirectly, the type of changes greens call for (Jacobs, 1995). According to Young, 'serious greens' are arguing not for the abolition of the state, but rather for its transformation (1992: 21).

The place of the state within green political theory should, I suggest, be assessed on its functional value in implementing the various green goals that together make up 'collective ecological management' and as judged by the criterion of ecological rationality. There is widespread agreement within the green political programme concerning the necessity and desirability of collective strategies for managing the environment. The debate within green political theory is essentially over the appropriate level (global, national or local), institutional form (state, market or community), procedural content (democratic, technocratic), normative composition (ecocentric, anthropocentric) and forms of knowledge (vernacular, local, scientific-universal) appropriate to collective ecological management. In the next section I outline an interpretation of this ecological management perspective as one way in which the 'greening of the state' can be fleshed out.

A Reconstructive Critique of Ecological Modernization

This section begins with a critical examination of attempts to 'green the machinery of government', focusing on what has become known as 'ecological modernization' (i.e. existing state-initiated environmental political practice). I then consider how this approach can be used to advance a more radical conception of green politics, collective ecological management, within which the state has a key role (i.e. possible environmental governance).

Although criticized for betraying an environmentalist or green reformist outlook, because it works within the existing institutions of modern societies, and does not seem receptive to anything other than an economic view of the non-human world, 'ecological modernization' does represent a 'realistic' theory of dealing with ecological problems and suggests one path for sustainable development. The transformations it demands, although not as radical as those proposed by eco-anarchists, are not as reformist and limited as critics often suppose.

The basic tenet of ecological modernization is that the zero-sum character of environment–economy trade-offs is more apparent than real.

Ecological modernization challenges the idea that improvements in environmental quality or the protection of nature are necessarily inimical to economic welfare, the fundamental position which dominated the early response to the 'environmental crisis'. In this earlier debate the green position was that a steady-state economy, in conjunction with zero population growth, was the only economy–ecology relationship which could ensure long-term sustainability (Daly, 1973b; 1985; Olson and Landsberg, 1975; Kerry-Smith, 1979).[7] In opposition to this idea, ecological modernization suggests that economic competitiveness is not incompatible with environmental protection; indeed, as Weale points out, 'environmental protection [is] a . . . potential source for future growth' (1992: 76). Future economic prospects increasingly depend on achieving and maintaining high standards of environmental protection. Key to this is separating economic growth from rising energy inputs (Dobson, 1995; Jacobs, 1996). In general terms then, ecological modernization can be viewed as an account of how existing political and economic institutions have responded to public pressure for governments to 'do something' about environmental problems. In terms of the ends/means distinction drawn above, it is concerned mainly (but not exclusively) with finding more sustainable means (technical, economic) to the same ends (continuing increases in material goods and services). As an ideology and approach to dealing with environmental problems it clearly originates from *within* the state system, rather than from within civil society, which is the usual source of green ideas. This is demonstrated by Weale's (1992: 75) analysis of it as a policy approach to pollution control, originating in a critical rejection of early policy approaches. Ecological modernization as an ideology is largely constituted by government programmes and policy styles and traditions, particularly those of Germany (1992: 79–85) and The Netherlands (1992: Chapter 5), and European Union environmental programmes, particularly the Fourth Environmental Action Programme (1992: 76–7). Thus one can say that the origins of ecological modernization lie in the environmental discourse of policy elites. However, seeing it purely as a legitimizing ideology for 'business as usual' would be a mistake, as argued below.

Ecological modernization for Weale (1992: 79) is understood both as a legitimating ideology within certain liberal states' response to environmental problems, and as a new departure in environmental policy principles. As such it can be viewed as marking a new environmental policy discourse from within the existing institutions of the liberal state – a form of institutional learning. Its emergence and strength as an ideology lie mainly in its capacity to render the imperative for economic growth compatible with the imperative to protect environmental quality. The evolution of this perspective has been described by Potier:

> By the mid-70s it had become clear that it is both environmentally and economically sound to anticipate the possible negative effects of an activity

> such as an industrial plant and to design it in such a way as to prevent pollution before it occurs. (1990: 69)

At the same time in the 1980s there developed a sizeable market for 'green' or 'environmentally friendly' products (Elkington and Burke, 1989). Finally there was greater public pressure for governments to tackle environmental problems (Weale, 1992: 167–70; Young, 1993: 53–6), as well as the legitimacy of government being increasingly tied up with providing environmental protection (Walker, 1989: 38; Weale, 1992: 1, 26). The congruence of these two factors, one from the demand side and the other from the supply side, represents the context within which ecological modernization developed. To use economic terminology, one could say that ecological modernization represents an 'equilibrium' policy position: a point at which supply (of ecologically friendly goods and services) and demand (for those goods and greater levels of environmental quality) meet. Thus it acts as an institutional (and ideological) compromise between economic and ecological interests, and as Weale suggests, a compromise between the economic imperative for capital accumulation and political legitimacy (1992: 89).[8]

On this view the state policy elites act as brokers and prime movers in encouraging interest groups, trade unions, industry, consumer groups and the environmental lobby to adopt and accept the agenda (and language) of ecological modernization. This character of the ecologically modernizing state – proactive, agenda-setting and interventionist – has led some commentators to view ecological modernization as an environmental neocorporatist political arrangement (Young, 1993: 88–90; 1994: 16–18). This is particularly so with regard to the centrality of some degree of economic planning within ecological modernization.

Ecological modernization is to be distinguished from radical accounts of 'sustainable development' and 'green economics', which argue that the compatibility of environmental protection and the economic imperatives is premised on the distinction between and separation of 'social development' and 'economic growth' (Eckersley, 1992b; World Commission on Environment and Development, 1987). Ecological modernization does not, for example, require alternative measurements of human welfare. Ecological modernization reconciles environmental and economic imperatives perhaps because it shares with the ideology of 'sustainable development' an essentially ambiguous character, which reconciles erstwhile opposing interests, i.e. 'environmental' ones on the one hand and 'industrial' ones on the other (Sachs, 1995; Richardson, 1997; Barry, 1998c). And while from an ideological green position this is a problem, from a practical political point of view it can be viewed as a positive advantage. Thus ecological modernization can be viewed as dealing with one aspect of social–environmental problems (between economy and ecology) from a state-centred 'administrative' perspective. It is not so much a political theory as an ideology of environmental public policy. However, despite its state-centred origins, ecological modernization can be used as a basis for a more radical position, which I outline in the next section.

For Weale:

> The challenge of ecological modernization extends beyond the economic point that a sound environment is a necessary condition for long-term prosperity and it comes to embrace changes in the relationships between the state, its citizens and private corporations, as well as changes in the relationship between states. (1992: 31–2)

As argued below, ecological modernization shades into collective ecological management when other considerations are posited alongside the institutions of ecological modernization. As Weale (1992: 78) argues, ecological modernization is in many respects a coping strategy adopted by states in the face of demands for higher environmental protection together with the continuing demand for conventional economic growth, which may become radicalized in terms of policy outcomes, when its focus shifts from dealing with the *effects* of environmental problems to their *causes*. Ecological modernization, particularly when placed within the context of a 'green welfarism' or green social democracy, may, like the emergence of the welfare state before it, be construed as an attempt to politically regulate production, via state planning and regulation, in response to the socialization of the (environmental) costs of production as a result of 'market failure'. For example, 'polluter pays' legislation, the precautionary principle, mandatory environmental impact assessments etc., all of which are central to ecological modernization, can be regarded as ways in which the environmental costs of production are either prevented or 'internalized' to some extent.[9] In this way, ecological modernization can be viewed as an ecological dimension to the modern state's 'crisis management' function.

Another way of putting it would be to say that ecological modernization becomes radical when the institutional focus moves from problem displacement to problem solution (Dryzek, 1987: 11). As Weale suggests:

> the stress upon setting emissions standards implies a focus on effects rather than causes and hence a focus on end-of-pipe solutions. We should not expect a great deal of emphasis, then, in policy discourse on the need for structural changes in production and consumption to reduce pollution. (1992: 84)

In other words, focusing on *ex ante* preventive measures, rather than *ex post* compensatory adjustments, may move one in the direction of structural change in the organization of the economy, as a necessary measure to deal with economic–ecological problems.

A problem with ecological modernization is that it does not go far enough in terms of the type, level and manner in which the changes it proposes are decided and implemented. That is, on its own, ecological modernization stands in danger of being a state-dominated and -initiated process of ecological management which is heavily weighted in favour of dominant 'industrial' interests. Ecological modernization can thus be viewed as an extension of the 'crisis management' function of the modern welfare state to include economy–environment relations. However, ecological modernization also, as Weale (1992: 150–1) suggests, implies a virtue-based conception of citizenship and some degree of public participation in environmental policy implementation. Nevertheless it is a very state-centred position. An example of ecological modernization's commitment to participation is given by Weale: 'An important aspect of project development is that there is typically a process of public consultation necessary before the project is allowed to proceed' (1992: 171). However in discussing public inquiries one must bear in mind that *whether* the proposed development should go ahead is not on the agenda; participation is limited to influencing the *manner* in which it proceeds. In other words, consultation and public participation within the ecological modernization model are restricted to discussing means not ends.

While not agreeing fully with Beck's assessment, I believe his critique of state-centred responses to ecological problems does bring out some of the basic problems with ecological modernization. According to him, 'The dangers of such an eco-orientated state interventionism can be derived from the parallels to the welfare state: *scientific authoritarianism* and an *excessive bureaucracy*' (1992: 230, emphasis in original). Ecological modernization remains concerned with finding means to specific ends and is unable to articulate the full range of normative issues relating to social–environmental affairs. When environmental problems become not just a matter of the most cost-effective legislative or market-based means by which they can be dealt with, environmental degradation may come to be seen not just as an economic externality requiring economic or scientific/technocratic answers, but also and primarily as normative issues requiring moral and political answers. That is, environmental problems are questions of 'right and wrong' and only subsequently matters of costs and benefits. At this point we may say that ecological modernization has shaded into the political-normative process of collective ecological management. That is, a transformation within the political regulation of social–environmental interaction has occurred. The mark of this transformation is the politicization (and not simply the bureaucratization or 'marketization') and moralization of environmental problems. While collective ecological management is different in kind from ecological modernization, this should not blind us to the areas of continuity between them. Within collective ecological management this requires radicalizing the institutional potentials of ecological modernization, particularly in respect to the transformation of the nation-state, and the relationships between state, market and civil society.

Towards Collective Ecological Management

Whereas ecological modernization, at least in part, is premised on a neocorporatist consensus between trade unions, the state, business, consumer groups and the formal environmental lobby, collective ecological management is not corporatist. It is not corporatist because a premium is placed on democratizing decision-making processes, as opposed to elite bargaining over policy. In opposition to elite bargaining we have popular participation through a variety of institutional mechanisms in which citizens are given more control over decision-making relating to social–environmental interaction and environmental policy-making. Thus collective ecological management can be viewed as a democratized and more radical form of ecological modernization. While there is a continuing role for the institutions of the nation-state, collective ecological management manifests a healthy degree of scepticism with regard to extensive state involvement, largely on the grounds that the state cannot and should not do everything. At the same time, collective ecological management favours decentralizing decision-making, *where appropriate*, to the local state level. Ecological modernization, on the whole, focuses on the national and international levels. Collective ecological management is not exclusively concerned with establishing ecologically efficient, i.e. materially productive, relations between economy and environment, and is state-based rather than state-centred. It is because ecological modernization takes as its subject of analysis the relationship between the economy and ecology, rather than society and ecology, that the state occupies a central role. Ecological modernization is about enhanced economic–environmental material exchanges in the name of continuing 'economic growth' or the 'competitiveness' of the national economy. Collective ecological management on the other hand, while not completely ruling out international trade, is concerned with shifting the orientation of the economy towards the national and local market, and views economic growth as one amongst other social goals. Its focus of concern is social–environmental relations and not simply economy–ecology interaction. Both ends and means are, within reason, open for democratic scrutiny and debate.

What stands out about ecological modernization is its constraining nature. There are at least three ways in which ecological modernization is constraining. Firstly, it focuses on elites, many of whom do not have any democratic mandate. Secondly, it limits the type of interests that can be brought to bear in establishing environmental policy. Those interests that can be articulated as economic values, conclusively established scientifically, or are matters of pressing political expediency are favoured. The onus of proof is on those who wish to protect the environment, rather than those who wish to develop it. Interests pertaining to the social metabolism with the non-human environment are reduced to political bargaining between elites and the assignation of economic values to environmental resources. This is related to a third limitation, namely, the fact that ecological modernization is premised largely on bargaining and trade-offs between vested interests. It is an interest-aggregating system rather than an interest-transforming system, where the interests articulated are a function of power, influence and access. 'Voice' is limited to powerful, established interests: hence the difficulty the environmental movement has had in gaining access to the policy-making process (Robinson, 1992; Aguilar, 1993; Jahn, 1993), and the division of the environmental movement into 'insider' and marginalized 'outsider' groups (Garner, 1996).

At root, ecological modernization works because the interests it balances are couched in the language of economic rationality. Environmental interests are considered only to the extent that these interests can be translated into the economic language of cost–benefit calculation. In order to protect the environment, it must first be demonstrated to be a 'resource', and preferably a resource with some direct and immediate economic benefits. In this manner one can say that the grammar of ecological modernization, and one of the main reasons for its success, is its use of neoclassical environmental economics. As indicated in the next chapter, neoclassical environmental economics reinforces not only the state-centred, bureaucratic nature of ecological modernization, but also the latter's choice of the market as the central organizing institution of the economy. The point is not that economic interests are not a legitimate way to value the environment, but that under ecological modernization it becomes the dominant form of valuation. This not only 'crowds out' non-economic forms of valuation, but more problematically obliges the latter to present themselves as economic forms of valuation. Against this view, the logic of collective ecological management in raising economic, ethical, scientific dimensions of the ecological crisis is close to Daly's view of 'green economics'. According to him, green economics '[relies] on market allocation of an aggregate resource throughput whose total is not set by the market, but rather fixed collectively on the basis of ecological criteria of sustainability and ethical criteria of stewardship' (1987: 7). Daly's emphasis on collective decision-making, identifying institutions (the market), and especially the combination of ethical and ecological criteria, is close to the aims of collective ecological management, viewed as a form of ecological stewardship.

However, what both ecological modernization and collective ecological management share is a concern with the political regulation of human interaction with the environment. Collective ecological management widens out the ecological modernization approach by viewing the totality of social–environmental relations as the appropriate context within which problems in the material interaction between economy and environment, as manifested in the 'formal' or money economy, can be placed.

In a sense the collective ecological management strategy is on one level a democratic political procedure within which various ways of valuing the environment (and thus various relations and interests to and in the environment) can be raised, deliberated and incorporated into policy recommendations. The green political case is thus not for a pre-emptive 'hands-off' approach. Rather, it recognizes that an adequate solution to environmental problems demands a wider context than that provided by economic or technical valuations of the non-human world. This can be taken to mean that in the case of social–environmental interaction, 'problem identification' is not to be left up to market processes alone. However, this is compatible with holding market processes as having a continuing role to play in the management or resolution of social–environmental problems. In other words, markets are often not the appropriate institutional setting for defining and thinking about social–environmental relations, though they are not ruled out as having some role to play.

Whereas ecological modernization places the onus of justification on those objecting to development, and a deep ecology view of green politics demands that the onus of justification be shifted to those who wish to use the environment (Dobson, 1990: 61), collective ecological management requires that there be no presupposition in favour of either preservation or development. Rather such issues must be resolved politically, which can be understood as implying that no form of environmental valuation or human interest in the world is exempt from public criticism and justification, particularly in the case of major land-use proposals, for example, road-building, mining and dam-building. The onus of justification falls equally on all who propose particular social–environmental relations. From a democratic point of view, we simply cannot say in advance whether 'use' or 'non-use' will be the outcome, although a public 'ethics of use', institutionalized perhaps in law or the constitution, serves to map the border between morally justified 'use' and illegitimate 'abuse'. That is, an ethics of use can function as a guide, an institutional representation of a society's collective moral assessment of *permissible* and *impermissible* uses of nature.[10]

The most important part of social–environmental interaction is the economy–ecology metabolism. To a large extent it is the *type* of governance pertaining to this metabolism which differentiates ecological modernization from collective ecological management, since both accept the necessity for such management. Collective ecological management differs in institutionalizing opportunities and fora in which social–environmental relations viewed in terms of right and wrong are not disadvantaged relative to positions in which they are viewed in terms of costs and benefits. This is what Beck (1992) means by 'institutionalized self-criticism'. It is not that 'moral' accounts of social–environmental relations are superior to 'economic' accounts, but rather that economic and technical accounts themselves embody normative presuppositions. And it is thus at the normative-political *not* at the technical level that these issues are to be worked out. Starting from the position that there are no value-neutral conceptions of the proper relation between society and environment, collective ecological management can be regarded as raising the normative character of different views of social–environmental relations to public and political debate. For example, whereas ecological modernization, largely underwritten by neoclassical environmental economics, seeks to monetize environmental preferences (which are exogenously given), and then calculate policy proposals based on information gained from their aggregation, collective ecological management focuses on the process of preference formation and transformation within deliberative fora. The making of environmental public policy decisions within collective ecological management is not a matter of aggregating pre-formed, given consumer 'environmental preferences'. Rather, the point is that the public good character of environmental issues requires a decision-making procedure in which individuals make judgements about the public good in question and not just private calculations. Attitudes to public goods are formed from given preferences when individuals are brought together within a public rather than a private setting. In other words, individual preferences are important as the starting point not the conclusion of decision-making about public goods. The appropriate setting for public goods is, as Elster (1986) suggests, a 'forum' rather than a 'market-place'.

The problem with aggregative approaches to social–environmental problems partly lies in the priority given to 'felt' over 'considered' preferences. As suggest earlier, decision-making based on economic preferences, while appropriate to private goods, is often not appropriate to decision-making in respect of public goods, such as the environment. After all, for the majority of social–environmental issues, the 'environment' in question is not a private good, although as indicated in the next chapter there are those who propose privatization as the appropriate solution. That is, environmental public policy decisions should not be made on the basis of aggregating preferences, but rather should be made on the basis of some conception of the common good. As critics of cost–benefit analysis (CBA) point out, there is a real problem in attempting to reduce individual values to preferences and make them commensurate by assigning them a monetary figure (O'Neill, 1993; Jacobs, 1997). Such preference-aggregating techniques, like market-based public decision-making, are not value-neutral. As Jacobs puts it, 'It encourages a particular approach to the valuation of environmental public goods, namely a "consumer" one in which private income is exchanged for personal benefit' (1997: 217). The point about a 'citizen valuation' approach is that the public good dimension is explicit. Faced with the classic choice between 'development' and 'environmental protection', the question that ought to be put to individuals is 'what should be done in the interests of society?' as a whole and not merely 'how much will it cost/benefit me?' Jacobs goes on to demonstrate the difference between 'citizen' and 'consumer' approaches to environmental issues by pointing out that in relation to the former question the appropriate payment vehicle is a tax, which all are required to pay, or more realistically, the shifting of tax revenues from one area of public policy to environmental protection. In other words, asking people how much additional tax they would be willing to pay to preserve the environment, knowing that the tax burden (cost) will be borne by society as a whole (i.e. which minimizes the free-rider problem), will generate qualitatively different answers than if they are asked how much of their own private income they would be willing to pay to preserve the environment (the standard CBA approach).[11] In other words, if the good is public, its costs should be borne by all, just as its benefits are available to all. Asking individuals as citizens to value environmental goods is thus markedly different from asking people as consumers to value them.

Part of this immanent critique involves the radicalizing and transformation of existing institutional responses to environmental problems, such as ecological modernization. As discussed in the next chapter, neoclassical environmental economics underpins much of ecological modernization. The former employs proxy-market techniques, such as contingent valuation exercises and cost–benefit analysis, which while based on aggregating the consumer preferences of a sample, are then used by state officials as the public or collective valuation of society as a whole. Once a sample has been collected in which individuals are asked to reveal their (hypothetical) consumer preferences in an equally hypothetical market situation, this is all the 'input' from 'society' that environmental economists and state officials require.

As has been suggested in previous chapters, the conception of green political theory being defended here does not imply *wholesale* cultural and metaphysical reconstruction. The resolution of the ecological crisis in a manner which would accord with green values and principles *does not* imply the ethical rejection of anthropocentrism, a central shared normative orientation of Western culture. Nor does it imply the political rejection of state institutions (as argued in this and last chapter), or the complete rejection of consumerism. But it does imply a process of *transformation* in these areas. These observations may be taken to back Goodin's statement that 'we can presumably live more or less in harmony with nature even in advanced industrial and post-industrial societies, without necessarily dropping out of those societies altogether', to which he added in a footnote, 'Indeed, the green political agenda would be even more pointless than its worst critic imagines were that not the case' (1992: 119). In other words, green political theory, conceived here as concerned with the political-normative process of institutionalizing an effective and ethically informed collective ecological management, works with the idea that part of this process involves the *immanent critique and transformation* as opposed to the *abolition* of existing institutional structures (eco-anarchism) and/or its existing cultural mediations of social–environmental relations (deep ecology).

That conceptions of nature have played a part as an identity-forming condition for human societies is undeniable. Nature in many respects is the original 'other' against which collective forms of identity are formed and maintained. As was indicated in the brief discussion on human nature and culture in Chapter 3, and in the analysis of bioregionalism in Chapter 4, particular understandings of nature and social–environmental relations are partly constitutive of a society's sense of itself and form a core aspect of its culture. This is a feature of all human societies from tribes to contemporary nation-states.

Attention to the cultural dimension underpins the suggestion made in the last chapter that the ecological crisis can be described as a *contradiction* within Western culture as opposed to a *total crisis* of that culture. Examples of this include Sagoff (1988) and Keat (1994) who argue that the environmental crisis can be viewed as a tension between the interests of individuals as consumers on the one hand and as citizens on the other. The resolution of this tension can only come from within the culture itself. In rethinking green politics, a fruitful and necessary step is seeing that there are resources within the present culture which can resolve its ethical and material environmental contradictions. As such the plausibility of the green message is dependent upon its ability either to 'tap into' the existing culture, or to extend or alter the existing one in new directions, not to reject it and create a completely new one. Thus part of the cultural aim of green politics involves the possibilities and necessity for transforming the existing national 'bioculture'. The 'bioculture' is defined by Taylor as 'that aspect of any human culture in which humans create and regulate the environment of living things and systematically exploit them for human benefit' (1986: 269–70). This is in keeping with the argument in Chapter 3 concerning the efficacy of adopting an 'inside-out' version of moral extensionism, rather than an 'outside-in' one based on some external critique and proposed 'new ecocentric ethic'. That is, start from existing social–environmental relations and their moral character and extend or transform the latter, rather than adopt a 'view from nowhere' or 'elsewhere' and propose sweeping changes in a moral idiom which is alien to the particular culture in question. Thus while cultures are not beyond criticism and change, any change to a culture must take place within bounds set by that culture.

The Land Ethic and Collective Ecological Management

According to Norton,

> if we are to break a new path, we must set our sights between artificiality – the choice to extinguish nature, to control it everywhere – and primitivism – the choice to isolate and save nature for its own sake: we must escape the environmentalist's dilemma by creating a culture that values nature independent of human demands, *but not independently of culture itself*. If there is to be a middle path, we must give up the idea that something must be either natural or artificial. Naturalness (wildness, too) must admit of degrees. (1991: 156, emphasis added)

In claiming that 'naturalness must admit of degrees', Norton's position is close to the position being defended here for collective management of the environment, and is compatible with an anthropocentric 'ethics of use'. This section focuses on Norton's work since it brings out the cultural dimensions of collective ecological management.

Although a much disputed topic, following Norton, one can find in Aldo Leopold's 'land ethic' an endorsement of collective ecological management. In Leopold's critique of the dominance of economistic thinking in determining social relations to the environment one can find strong intimations of what Norton terms an 'integrated theory of environmental management' (1991: Chapter 3). One reason for drawing on this interpretation of what many deep ecologists (wrongly, as argued later) regard as the first modern expression of their position is that Leopold was critical of the way in which economic valuation of the natural world 'crowded out' alternative valuations, i.e. social, scientific, aesthetic, ethical. However, one must stress that Leopold did not reject economic interests in nature so much as to put them in their proper context and prevent them from dominating social relations to the land. A clear expression of this is his statement:

> The 'key-log' which must be moved to release the evolutionary process of an ethic is simply this: quit thinking about decent land-use as solely an economic problem. *Examine each question in terms of what is ethically and esthetically right as well as what is economically expedient*. A thing is right when it tends to preserve the integrity, stability, and beauty of the biotic community. It is wrong when it tends otherwise. *It of course goes without saying that economic feasibility limits the tether of what can or cannot be done for land. It always has and it always will*. (Leopold, 1968: 224, emphasis added)

Unfortunately, it is the text sandwiched between the italicized sentences that has received most attention in green discussions of Leopold's 'land ethic'. Placing his famous statement within context, Leopold's position has less in common with deep ecology than is generally assumed.

Leopold's 'land ethic' did not envisage a 'hands-off' approach to social–environmental relations of the type favoured by most deep ecologists. According to Norton, 'Leopold never questioned the *right* of humans to manage [the environment]: he questioned, rather, our *ability* to manage' (1991: 53, emphasis in original). Equally significant is Leopold's suggestion to 'examine each question in terms of what is ethically and esthetically right *as well* as what is economically expedient'. In other words, economic issues are not to be abandoned, but rather economic valuations are one important set alongside questions of ecological sustainability and cultural valuations of the environment.

In agreement with Norton, we ought not to see Leopold's 'land ethic' as a prototype deep ecology statement of the 'intrinsic value' of the non-human world. Rather we should interpret it as an attempt to develop a theory of collective ecological management, in which culture plays a central role. That is, the land ethic holds that the values of nature are culturally defined, and cannot be independent of that cultural context. As Norton puts it, 'Leopold saw the search for such an ethic as one culture's search for a workable, adaptive approach to living with the land' (1991: 58). The land ethic, although it does argue for harmony between society and its environment, begins from an environmental management, or stewardship, perspective which recognizes that harmony is created rather than discovered or given. The social construction of this harmony is the aim of collective environmental management, and the aim of green politics is to argue for an integrated form of environmental management in which aesthetic, scientific, normative, democratic as well as economic considerations can be brought to bear. The aim is to find the best 'adaptive fit' between particular cultures and their environments.[12]

As such, collective ecological management and ecological culture are to be seen as part of a process by which the changing relationship between culture and environment can be brought to an equilibrium position. However *there is no fixed point, but a variety of possible equilibrium positions*. That is, for green politics *qua* collective ecological management, there is no static social–environmental metabolism. Equilibrium or harmony is dynamic not static. The aim of green politics is obviously to encourage sustainable and symbiotic forms of social–environmental interaction. The latter is a metabolism which seeks to ensure that both human and non-human interests are made as compatible as possible, understood as maximizing long-term human interests compatible with minimizing human environmental impact.

In ecological terms what environmental management implies is human intervention in natural systems which aims for what Odum calls an 'anthropogenic sub-climax' (1983: 473). This is an ecological state which is different from that which would have been attained in the absence of human intervention. In other words, this sub-climax state is the 'humanized environment' mentioned in Chapter 3. The range of sub-climax states thus specifies the range within which stewardship is operative. This 'humanized environment' is not an artificial environment, but rather an environment transformed by and in part the combined outcome of human intentional activity and natural processes. Human intervention prevents the 'natural succession' of ecosystems to climax states, instead transforming and maintaining ecosystems in accordance with human interests (which can of course include a concern for the interests of non-human entities). An indication of the political significance of this is given by studies in the UK which show that 'public opinion favours a conserved and planned landscape, rather than the one allowed to revert to a more natural, climax state' (Pearce et al., 1993: 107–8). In the UK context, the issue is not whether natural climax states ought to be permitted to emerge because they are somehow morally superior (because they are 'natural') to socially managed sub-climax states, but rather which particular sub-climax state to choose, given that there is no simplistic identification of the 'natural' and the 'good'. In terms of the distinction often used in ecological debates, the British environment, and that of Western Europe more generally, is more of a 'garden' (i.e. a humanized environment) than a 'wilderness'. And one of the reasons why British opinion favours a conserved landscape may have something to do with the fact that such landscapes bear the mark of previous human transformation and thus constitute a living connection with the past. That is, the preference for preserving such humanized landscapes can partly be explained by their importance in the constitution of collective identity, the 'heritage' of the past. 'Landscape', denoting a humanized environment, can be viewed as a 'text' upon which is recorded the efforts of those who came before us. Indeed, it is worth noting that it is difficult to see how 'place' can become an important identity-forming category without its transformation into 'landscape' and/or its incorporation into an understanding of the environment as 'homeland'. That is, preservation of the 'landscape' acts as a permanent reminder that the creation of a successfully humanized environment (the 'sub-climax' state that is our ecological niche) is not the work of the present generation alone, but the product of many previous generations. Thus in calling for the preservation of such humanized landscapes one is also calling for the preservation of cultural heritage.[13]

Allowing natural ecological succession to resume within the context of an environment that has been intensively managed for hundreds of years would not only cause extensive and unnecessary social disruption and upheaval, but also cause suffering or harm to those natural entities which have carved out niches within the humanized ecosystem. Natural succession and the attainment of a natural climax state is only one possible equilibrium between society and its environment. In terms of ecological rationality, there are other possible equilibria which do not require the 'quietism' involved in the prioritization of natural over human-induced ecological succession, or any moral priority to these states. According to Milton, favouring 'natural climax states' as the template or ideal for social–environmental harmony requires a particular type of environment, namely a 'giving' or abundant environment (usually associated with a hunter-gatherer social form), rather than a 'reciprocating environment' which requires human intentional transformation in order for humans to survive and prosper (1996: 118–19). 'Ecological climax states' can be taken as the polar opposite to the complete human management of the environment to the point of replicating its functions within the 'technosphere'. On this account, collective ecological management is located between 'non-interference' and 'substitution', or to recall Norton's words, between 'primitivism' and 'artificiality'. The advantage of this view of the 'land ethic' is that 'ecosystem health' and integrity *includes* human activities which do not threaten the critical threshold levels of the chosen 'anthropogenic sub-climax'.

The cultural dimension of green politics *qua* collective ecological management can also be explained in reference to forms of knowledge. Similar to the green principle of decentralization, understood as meaning that nothing should be done at a higher level if it can be done at a lower one, it is also the case that different forms of knowledge may be more appropriate as we move from global to national to local levels of social–environmental relations. At the same time, 'The world needs a variety of sustainable societies achieved by many different paths' (IUNC et al., 1991: 8). At the global level normative concern for 'nature' as a whole, or for whole species, is rather abstract. As Rolston notes, 'A duty to a species is more like being responsible to a cause than to a person' (1988: 144). Caring for the earth involves a different epistemological framework for making ethical judgements than caring for a particular animal or a particular environment. Corresponding to this abstract ethical commitment, at the global level of 'thinking about nature' (Brennan, 1988), it is likely that ecological science may be the most appropriate form of knowledge at this level, and upon which transnational agreement may be reached as how to deal with global environmental problems. As we move down to the national level, ethical concern becomes more culture-specific as 'nature' becomes 'the land' or a determinate environment. At these levels the modes of interaction appropriate to regulating human action, while still having a place for science, are likely to have a 'thicker' normative texture. That is, interacting with a determinate and familiar aspect of the environment is more likely to involve strong and concrete moral commitments. This is particularly so if one takes sympathy as an essential component of moralizing social–environmental dealings as suggested in Chapter 3. The moral experience of nature is richer since it is more partial at the level of particular, familiar parts of the environment.

This is because we *experience* nature in the particular, rather than in the abstract. We *perceive* the local environment, but *conceive* of the global one. Knowledge of the world gained from direct experience of determinate parts of the environment will have a different effect on our ethical treatment of the non-human world than will abstract knowledge of that world. However as Plumwood points out:

> Special relationships with, care for, or empathy with particular aspects of nature as experienced, rather than with nature as abstraction, are essential to provide a depth of concern . . . experience of and care and responsibility for particular animals, trees, rivers, places and ecosystems which are known well . . . enhance rather than hinder a wider, more generalised concern for the wider global environment. (1993: 187)

It is interesting to note that both moral sympathy and science can be understood as universal modes of human experience. The difference between them is that science is universal and impartial, sympathy is universal but partial. That is, sympathy is a natural feature of what it means to be human, which makes it universal. However, sympathy is directed largely to partial rather than impartial objects of concern. Stewardship as a virtuous mode of interaction, a social practice with internal standards of excellence, may be said to be a combination of sympathy and science. Sympathy without knowledge may lapse into 'sentimentality', while knowledge without sympathy may lead to an 'arrogant humanism'. Both are ecological vices to be avoided.

The importance of science (as a non-local form of knowledge) in the establishment of effective collective management strategies cannot be overestimated. Not only is it the case that science is required in the elaboration of effective solutions to national environmental problems but also scientific understanding is indispensable in dealing with transnational and global environmental problems. Without going into an analysis of the sociology and politics of science, and aware that 'science' is not a monolithic body of agreed knowledge, and that it can be used to support privileged groups (Sachs, 1995; Thompson, 1995), all I wish to suggest here is the potential role scientific discourse can play as a common language for dealing with international and global environmental problems. It can also function as a shared metaphysical basis for green politics in conjunction with Norton's (1991) suggestion that science can create the basis for practical agreement between ecocentric and anthropocentric positions.

At the local or micro-level we arrive at ethically richer and stronger attachments to parks, gardens, allotments, forests, watersheds, mountains, particular animals and species.[14] At the local level more vernacular idioms of knowledge relating to ecological management may be appropriate. This emphasis on the importance of the vernacular has been a constant theme within green politics beginning with Illich's critique of the dominant Western model of development as the 'modernization of poverty' (1974) and the destruction of local, autonomous social practices as a result of the rise of 'disabling professions' (1977). In more recent years, the defence of the vernacular, the local and the traditional, has been a dominant theme within Southern green politics as a response to globalization (Shiva, 1988). These issues are dealt with in more detail in the next chapter.

As we move from the local to the global, ethical concern for the non-human world becomes more abstract, while the forms of knowledge appropriate to the regulation of human interaction with that world become more 'objective', universalistic and impartial. But without the meso-level of ecological culture, situated between the macro-level of the global biosphere (where 'humanity' interacts with 'nature') and the micro-level of particular environments (where individuals and communities interact with particular animals, parks, urban spaces), the connection between them would be to a large degree opaque culturally and politically speaking. It is through the shared (and contested) grammar of culture that the ecological narratives of the local and the global may be created and transmitted while individuals can place themselves within wider and wider ecological contexts. In this sense the green political project as given by collective ecological management involves arguing not for a 'return to the land' in some pre-modern sense of recovering a lost rural and more harmonious and 'natural' way of life (Ferris, 1992), but rather for the reintegration of the 'land' both ethically and ecologically into Western cultures.

Given the present global reach of human activities (which is not to say that *all* humans are responsible), a full understanding of the metabolic interaction between human societies and the non-human world requires placing them in relation to ecosystemic levels from their local 'ecological hinterland' and through more remote ecosystems right up to the global biospheric level. The majority of presently existing societies interact with the non-human world on both 'ecosystemic' as well as 'biospheric' levels. The latter can be taken to mean that the 'human impact' at this biospheric level is the sum total of global national interactions, that is, the aggregate 'total biospheric impact' of human metabolic activities organized on a national basis.[15] It is to the particular forms of organization for collective ecological management that we turn next.

Ecological Management, Planning and Governance

For many writers on green issues, it seems almost inevitable that dealing effectively with environmental problems requires institutional responses in which there is some degree of planning. This is particularly so, as one would expect, with eco-socialist positions (Stretton, 1976; Mulberg, 1992; 1993). It is also an integral part of green social democratic proposals (de

Geus, 1991; Eckersley, 1992a) and ecological modernization (Weale, 1992; Young, 1994). In the last few years national environmental plans have been drawn up across a number of Western democracies from the Canadian Green Plan (Selman, 1994) and the Dutch National Environmental Policy Plan (Van der Straaten, 1992) to the Norwegian Samla Plan for the management of water resources (Rothenberg, 1992). At the same time issues around 'environmental management' have been at the heart of discussions attempting to spell out the policy implications of sustainable development (Pearce et al., 1989; Carley and Christie, 1992). While the adoption of a planned approach to ecological management is often an indication of the level of public concern, and the extent of social support and commitment to protecting the environment, there are a variety of other reasons why some degree of public planning or 'steering' (Jacobs, 1995) is an essential aspect of green politics.

The attraction and necessity of planning and co-ordination can be seen as partly to do with the search for *integrated policy approaches* to dealing with environmental problems (Weale, 1992: 93–118). Following Dryzek (1987: 10–11), the problem is that a purely political or economic 'solution' to an environmental problem more often than not simply *displaces* it rather than solves it. Displacement can cross media (water pollution to solid waste pollution), in space (exporting pollution) or time (to future generations), from one institution to another (externalities from the economy being passed to the political sphere) or within an institution (from one department to another). There is thus a gap between these 'solutions' to ecological problems and ecological solutions. Thus while it may be economically or politically rational, in the short term, to pollute the environment in excess of its ability to assimilate that level of pollution, it is clearly not ecologically rational in terms of the environment's capacity to sustain non-decreasing levels of human (and non-human) welfare. Norton (1991) argues that the character of many ecological problems is such that they demand an integrated, holistic and flexible problem-solving approach. That is, the *solution* as opposed to the *displacement* of pollution problems, for example, requires some degree of public regulation and planning.

Planning within the context of collective ecological management can be thought of as analogous to collective choice under conditions of uncertainty. Planning within this context is more about the maintenance of a process than the achievement of any particular outcome. Given the uncertainty which characterizes environmental problems, any environmental planning will necessarily have to be flexible, be sensitive to ecological feedback and not cause irreversible ecological changes. To use Beck's (1992) terminology, what is required is a 'reflexive' and flexible form of collective regulation understood as a social learning process as well as an exercise in social–ecological adaptation. This is what Eckersley means when she claims that 'The task of integrating environmental and economic decision-making requires a new (and yet to be developed) administrative framework that is considerably more flexible, collegial and consultative than the traditional model' (1993a: 22). This can also be understood from an ecological modernization perspective, which uses a modified version of the traditional model, but with the recognition that 'If national planning is shunned, its alternative will have to be invented' (Weale, 1992: 149).

An important ecological benefit of state regulation, which would be enhanced via planning, is its efficacy in securing environmental innovation in production. That is, one of the failings of the ecological modernization approach is its tendency to rely heavily on 'end of pipe' solutions to pollution instead of producing genuine innovations in production to prevent pollution in the first place. There is empirical evidence that regulation acts to encourage environmental innovation (Ashford, 1993). In this way there is some evidence that state regulation and planning can stimulate environmental invention and innovation.

As used within the context of ecological governance, planning refers to conscious, collective determination and co-ordination of social activities in the pursuit of the maintenance of some chosen rate of metabolic exchange. Planning therefore does not refer to social reorganization in accordance with some state-imposed blueprint. That is, the planning process within collective ecological management is not about wholesale 'social engineering' (Popper, 1974: 158–67). But nor is it 'piecemeal', implying discrete, *ad hoc* policy change. Some, like Jacobs (1995), have sought to see the issue not in terms of planning versus the market since the achievement of sustainability requires governmental intervention, particularly with regard to the economic sphere. As he puts it, 'The opposition between "market mechanisms" and "regulations" is overstated, missing the common requirement for hands-on government intervention' (1995: 1).[16] It would be more accurate to describe environmental planning as involving a process of 'ecological restructuring', altering and reintegrating the relationships between state, society, economy and environment on ecologically altered principles (de Geus, 1996). This macro-level, institutional change parallels the micro-level change the virtues of stewardship seek to realize in integrating the different roles individuals and groups have as citizens, parents, producers and consumers.

An example of the role of planning and the state within ecological governance is the precautionary principle. The precautionary principle, together with the 'polluter pays' principle, is one of the most important green policy principles. It is a principle of prudence, particularly under conditions of uncertainty, which makes it particularly suitable for environmental issues. As Boehmer-Christiansen (1994) argues, the precautionary principle evolved out of the German socio-legal tradition and was explicitly set within the context of state management and planning. Thus the precautionary principle is an interventionist measure, a justification of state involvement in the economy in the name of good government (de Geus, 1996). It thus has much in common with the interpretation of ecological modernization given by Weale (1992), which is unsurprising since both can be placed within the German *Rechtsstaat* tradition. The significance of the precautionary principle lies in the fact that it serves to justify state management of resource use because one of the central aspects of this principle is the idea that the onus of proof is on those who propose major social–environmental change. The operation of the precautionary principle makes the state and its agencies the guarantor of the environmental status quo. This principle expresses in a modern legal form the virtue of prudence in respect to development demands on the environment. Although the precautionary principle seems to reintroduce the rejected idea of a 'hands-off' policy regulating social–environmental exchanges, this is not the case, although it does strengthen the case for environmental preservation when compared with the current deregulation of social–environmental interaction. The precautionary principle and the role of the state in implementing it, typically within the context of environment versus development disputes, stress the need for discrimination and careful consideration when faced with development proposals. Thus the precautionary principle is not anti-development *per se,* nor is it pro-preservation, since development is always possible so long as it is in accordance with the precautionary principle. However, the main point here is that solving environmental problems as part of a collective ecological management strategy requires such *ex ante* policy principles as the precautionary one, an aim of which is to possibly prevent environmental problems from arising in the first place. And such *ex ante,* preventive measures require some degree of democratic public management, accountability, consent and state enforcement.

Another reason for some degree of conscious planning and management comes from the idea that the standard green reading of ecology which sees ecosystems as self-regulating and harmonious is a disputed claim within ecological science. As Brennan points out:

> the whole issue of whether ecosystems are generally self-maintaining diverse systems or simply fortuitous groupings of populations that at least for a time are not lethal for one another is very much undecided in ecology. It is striking, and unfortunate, that many conservationists operate with ideas of balance and diversity in nature that were more prevalent in the nineteenth century than among contemporary ecologists. (1992: 16–17)

As indicated in the last section, maintaining ecological diversity within social–environmental relations requires active social intervention.

Added to this is the controversy over the whole idea that diversity and ecosystem complexity are positively related to ecosystem stability. According to Clark, and contrary to popular perception, 'The highest diversity of species tends to occur not in the most stable systems, but in those subject to constant disturbance, e.g. rainforests subject to destruction by storm, and rocky intertidal regions buffeted by heavy surf' (1992: 42). One implication of this is that if biodiversity is a desired value this may imply human intervention in ecosystems to maintain high degrees of biodiversity, perhaps by disturbing ecosystems in the requisite manner.[17] Of the many conclusions one can draw from this perhaps the most important is that a 'hands-off' approach may not guarantee the types of ecosystems that many deep greens desire. Ecosystems characterized by diversity, balance and complexity may have to be actively created and managed. The normative standpoint from which to view social impacts on the environment is not a 'hands-off' position which frames the issue of the relation between society and environment in terms of 'use' and 'non-use', but rather that proposed in the last chapter as the 'ethics of use' which attempts to distinguish 'use' from 'abuse'. Once this ethical issue has been settled, ecological science can then be used to help distinguish 'good' from 'bad' ecological management, i.e. sustainability from unsustainability. The type of social regulation implied by collective ecological management involves the normative constraining of permissible policy options. We could imagine this as the democratic character of collective ecological management. It is not that each social–environmental issue is to be dealt with by all citizens taking a vote on the issue; rather it is that citizens (as opposed to bureaucrats and experts) can participate in what Jacobs (1996: 13) calls 'decision-recommending' rather than decision-making institutions. That is, such forms of popular democratic participation lay out the parameters of 'use' and 'abuse' in particular cases, that is, the normative bounds of environmental management which can then be carried out through state institutions.

In this way the idea of political ecological management begins with normative issues and then moves to employ ecological science, and other more local and particular forms of knowledge where appropriate, to help in the formulation of an effective strategy for managing the ecological commons.[18] From the point of view of collective ecological management, planning and management will be informed by the cultural valuations of nature of the society in which it operates. A good example of this is presented by Rothenberg (1992) who contrasts the Norwegian Samla Plan with the American Endangered Species Act as expressing two broadly similiar ecological policies, but underwritten by two different cultural valuations of the environment. Whereas the Samla Plan (a management plan for Norway's water resources) assesses costs and benefits of damming rivers in a manner which is sensitive to the claims of ecological communities of wildlife and ecosystems, the US Endangered Species Act is individualistic in identifying particular species rather than ecological communities as the object of environmental management and legislation. Rothenberg argues that this individualistic–community difference is due to the different cultural contexts of the two countries. Tellingly he notes that

> One might argue that the Samla Plan only works because the ideal of Norway is sufficiently unified in people's minds to agree that each part of the society needs to compromise to serve the whole. *The belief that a common goal is important makes the system singular, not pluralistic.* Perhaps the individually centred American system is even more pluralistic, as it admits no common ground save 'don't tread on me!' – respect for the rights of other individuals waiting to be conceived and tagged. (1992: 131, emphasis added)

The sensitivity to ecological communities can be interpreted as an extension of Norwegian political culture in which community and the idea of the environment as an overarching national collective good are central. Such a political culture is lacking in multicultural, pluralistic America; hence the individualistic ethos of its environmental legislation, where concern is not for the 'national environment' as such but for particular threatened individual species or ecosystems.

However, though state planning and co-ordination may be invaluable in dealing with *ad hoc* contingencies, such as oil spills, it is unlikely that hierarchical bureaucracies will be completely effective in dealing with the complexity and uncertainty associated with all environmental problems. As Dryzek notes, 'Highly structured organizations are at a loss, though, when it comes to dealing with high degrees of uncertainty, variability, and complexity – circumstances that are, of course, ubiquitous in the ecological realm' (1987: 108). In other words, a centralized, bureaucratic organization such as that associated with the contemporary nation-state may be ecologically rational only across a specific range of environmental problems. This is an argument not so much against the state *per se*, but against centralized, hierarchical and bureaucratized planning and administration in some cases where this type of governance is not appropriate. An alternative conception of the state, that of a flexible, enabling state with minimal centralization and bureaucratization, is one that concurs with the goals of collective ecological management. Hence the focus on the local state as the appropriate level for sustainability in Agenda 21 of the Rio Declaration, which will be discussed in the next chapter. It is not that the state manages society's metabolism with nature on society's behalf (the extension of the welfare state to the ecological sphere) but rather that a democratized state is a necessary institution which, in conjunction with other non-state forms of governance, contributes to the overall goal of collective ecological management. If we widen out the context of solving environmental problems as a matter of governance then the state may be less ecologically irrational in its acts since it becomes one institutional aspect of a much wider social process. Within the context of ecological governance the state does not stand completely condemned on ecological grounds, particularly if flexible forms of administration and decentralized forms of co-ordination, control and feedback are employed.

Ecological Restoration

Whereas the majority of environmental problems with which collective ecological management has to deal concern direct ecological inputs to human welfare, such as ecological resources necessary to economic activity, there is another type of society–nature relation which equally calls for large-scale social organization and co-ordination. While most environmental problems have to do with immediate supply problems or discrete ecological goods and services, 'ecological restoration' has to do with repairing degraded ecosystems or landscapes. In some cases this involves returning a disturbed ecosystem to the status quo ante, that is, either its stage of ecological succession prior to human interference, or its ecological state as a 'humanized environment' prior to industrial forms of human interaction. This understanding of restoration ecology involves, according to Morrison, 'returning a site to some previous state, with the species richness and diversity and physical, biological and aesthetic characteristics of that site before human settlement and the accompanying disturbances' (1987: 160). However, one can also posit understandings of ecological restoration as concerned with returning an ecological area to a past anthropogenic sub-climax state, that is returning a site to a 'pre-modern' or 'pre-industrial' state. The basic premise of ecological restoration is that human interference can, to some extent, cure the wounds it inflicts. The novelty of ecological restoration lies in that it is an example of social–environmental interaction where the environment is dependent upon social agency, rather than in the non-reciprocal manner in which relations between society and environment are usually understood.

The large and growing philosophical debate and controversy around ecological restoration will not be entered into here.[19] One point that may be raised concerns the fact that ecological restoration, perhaps more than any other social–environmental relation, highlights the potentially symbiotic (and sustainable) as opposed to parasitic quality of social interaction with the environment. In the case of restored ecosystems, and to a lesser degree in other managed ecosystems, there is a mutual reciprocity in the metabolic relation that is often missed in talk of the non-reciprocal dependence of humans upon nature. Restored ecosystems are a stark reminder that there are situations where the dependence is reversed, i.e. where nature is dependent upon society. That is, one can view ecological restoration as another type of environmental management. Of course this 'nature', or more technically the particular ecological succession state achieved by social intervention, is human-determined.

However, this does not mean it does not partake of 'naturalness'. Recalling Norton's statement 'naturalness admits of degrees'.

For present purposes it is assumed that the restoration of degraded ecosystems and landscapes is both possible and desirable, perhaps even necessary.[20] The main point I wish to make concerning ecological restoration is that on both these counts, its desirability and its possibility, it requires large-scale social organization, regulation and intervention. It is because of the costs involved in ecosystem rehabilitation that the state and its agencies are often the appropriate institutions to carry it out. The state with its tax revenues and its administrative and co-ordinating ability, together with its access to the relevant ecological knowledge (not necessarily that given by ecological science) and its legitimacy, is clearly in a strong position to restore ecosystem health. Indeed, it may have a statutory duty to restore ecosystems as part of fulfilling its obligations under both national and international law, to preserve and increase domestic biodiversity levels for example.

It may also be that repairing environmental damage requires institutions other than those appropriate to environmental preservation. Extending the subsidiarity idea that decisions ought to be taken at the lowest possible level, we may also add that institutions should be relative to the particular issue at hand. It is because of the range and type of social–environmental issues that ecological governance deals with that no one institution, such as the centralized agencies of the state, will be sufficient to cope with them all. That is, the various social–environmental interactions and relations that together make up the totality of a society's metabolism with its environment demand different institutional forms, principles and modes of operation. And while I hope to have demonstrated why the state will still have a role to play within collective ecological management, I also hope to have shown that the latter can only be understood in terms of 'governance' rather than government. The centralized state cannot and should not do everything, and the emphasis on 'governance', as opposed to 'government', is to make this explicit.

Conclusion

The main question this chapter has addressed is the issue of the most appropriate institutional structure for green politics. It amounts to an admittedly incomplete defence of a form of green politics which seems to break with most of the touchstones which are used to determine whether or not a theory is 'really' green. Firstly, it is anthropocentric, but this is a self-reflexive anthropocentrism which does not suppose that any human reason, simply by being human, justifies any use of nature. Secondly, and following on from the latter, it pitches the political-normative debate at the level of an immanent critique of Western culture. Leopold's 'land ethic', for example, does not require widespread metaphysical reconstruction. Thirdly, it does not reject economic valuations but rather seeks to integrate them within a wider political process of environmental valuation. Finally, it is committed to the state as a key institutional feature for the governance of a sustainable and symbiotic metabolism with the non-human world.

Collective ecological management is not simply about institutional changes, but is at its heart a normative political process for deciding on a social–environmental metabolism, which, in part, is to choose to live in a different sort of society. Collective ecological management is *not* about state management of the environmental affairs of civil society, that is the state exclusively 'taking care' of the public issues of policy-making and implementation. Ecological management as understood here includes but extends this image to encompass forms of collective action, deliberation and implementation which go beyond existing liberal democratic theory and practice. Issues of the democratic character of collective ecological management are dealt with in Chapter 7. However, before moving on to discuss the relationship between democracy and green politics, we examine some central political economy issues as part of the task of rethinking green political theory.

Notes

1 As mentioned in the last chapter, realistic eco-anarchistic theories do not suppose that non-state forms of governance are non-coercive; they simply require different, less formal forms of coercion, as Taylor (1982) points out. An example of 'decentralized restraint', what O'Riordan (1981: 306) classifies as an 'authoritarian communal' solution, can be found in Goldsmith (1972b) and Heilbroner (1980).

2 One way of looking at it would be to argue that eco-anarchistic values may be expressed through social practices supported by state institutions. That is, they could be *honoured* as opposed to *realized*. What is meant by this is that the values and practices of eco-anarchism, including *inter alia* communal self-determination, care for a particular environment, liberty, equality and well-being, may be *represented within and influence* the institutional structures of

society. In short, honouring eco-anarchistic values does not involve the wholesale rejection of state institutions as would be the case if these particular values and practices were to be realized. Eco-anarchistic values and practices are, after all, simply *alternative conceptions* of green values and practices. This is an implication of seeing eco-anarchism as a regulative as opposed to a constitutive ideal.

3 One of the most important social practices from this point of view is citizenship, the outlines of which will be presented below and developed further in Chapter 7. Other social practices which impinge directly on social–environmental transformative activities include science and technology as well as more direct forms of interaction such as farming, animal husbandry, forestry and work. Work is discussed in more detail in Chapter 7.

4 As argued in the next chapter, the elevation of sufficiency as a principle regulating the economy–ecology metabolism necessitates the re-embedding of the economy within society in Latouche's (1993) terms, understood as the collective self-conscious limiting of economic imperatives. Sufficiency as a culturally and politically defined category represents an important feature of collective ecological management, namely as the process by which society imposes limits upon itself. The clash between maximization and sufficiency can be readily observed in the debates over competing conceptions of development and progress between green critiques of economic growth and competing visions of 'sustainable development'. Another aspect of this has to do with the value of self-reliance and sufficiency as ecological virtues.

5 An 'anthropogenic sub-climax' (e.g. an agricultural ecosystem as against a wild/unmanaged ecosystem) is an ecological state which is different from that which would have been the case in the absence of human intervention. One way of understanding this concept is that this anthropogenic sub-climatic state relates to the 'humanized environment' which constitutes our 'ecological niche'. It can be considered as the outcome of human ecological management. Thus, it is only in the absence of human interests that it makes sense to argue, as Commoner (1971) does, that 'nature knows best'.

6 More contemporary analyses of the distributive implications of green politics also emphasize that any putative 'green theory of social justice' must take into account that what is to be fairly distributed includes social and environmental costs or risks as well as environmental and social benefits. See for example Beck (1992).

7 It is worth noting that although the steady-state economy position is at odds with the underlying principles of ecological modernization, both agree on the importance of an interventionist, proactive state. For example, Daly (1987), a leading proponent of the steady-state economy, argues that the advent of a society premised on the minimization rather than the maximization of 'economic throughput' has as a corollary the need for the state to manage ecological resources and distribute the available wealth. The latter justification for the state comes from his argument, with which most green theorists and activists would agree, that economic growth is a substitute for tackling the problem of socio-economic inequality. To limit inequalities he has proposed 'a minimum income coupled with a maximum income and a maximum on wealth – a limited band of inequalities necessary for incentives, for rewarding work of varying irksomeness and intensity, yet ruling out extreme inequality' (1985: 125).

8 In this respect, ecological modernization as an ideology at the national/regional level is similar, in terms of both its origin and its function, to 'sustainable development' at the international level. Both depend for their success on rephrasing the economy–ecology debate in a language which harmonizes ecological sustainability with orthodox economic imperatives. This is done by translating the environment into the language of orthodox neoclassical economic theory (Pearce et al., 1989; 1993) so that the environment is recast as an 'economic' problem which dissolves its political-normative dimensions. Both ecological modernization and sustainable development are state- or supra-state-level responses, and thus find their origins in the policy discourse of bureaucratic management. See Barry (1998c), Dryzek (1995; 1997), Sachs (1995) and Richardson (1997).

9 For some writers on ecological modernization this 'internalization' process goes beyond purely economic considerations to more political notions of self-regulation and 'governmentality' (Neale, 1997). However, there is empirical evidence to suggest that the rate and effectiveness of environmental innovation are enhanced under the regulatory regime which is characteristic of ecological modernization. That is, the planning and regulatory approach of ecological modernization may deliver environmental innovation in a way voluntary arrangements cannot.

10 While there are similarities between a public 'ethics of use' and certain understandings of justice, it is not my view that such an ethics of use should be thought of as a green theory of justice which extends considerations of justice to the non-human world. A more appropriate view of the ethics of use would be to see it as a legal side-constraint on social–environmental interaction, perhaps institutionalized within the constitution of a democratic polity. That is, the law can, and already does, express collective moral judgements which are not, strictly speaking, considerations of justice. For example laws against cruelty to animals are not about being 'just' or 'unjust' to them. If our relations to the non-human world are not matters of justice, but nevertheless require moral deliberation as green politics suggests, then it may be that the most appropriate institutional form for an ethics of use is a legal one. See Barry (1995d) and Eckersley (1996a) for an examination of this legal option, and Norton (1994) and Saward (1996) for constitutional interpretations.

11 It is also likely that reversing the question, and asking how much individuals would require to be adequately compensated for the loss of some environmental amenity, would lead to monetary amounts far in excess of those arising when they are asked how much they would be willing to pay to prevent such a loss. The discrepancy between *ex ante* and *ex post* amounts may be due to self-interest, since compensation is paid by someone else, the public authority. In other words, compensatory approaches encourage 'free-riding' since the costs are borne by all, but the benefits are enjoyed by the individual. However, the same compensatory approach applied to citizen as opposed to consumer decision-making would not have the same free-riding problems since the appropriate compensation might be either a decrease in the tax rate, or an increase in welfare benefits, or perhaps a public holiday or festival or some other public benefit. Asked as a citizen, the individual knows that the compensation will be shared by all.

12 Following Milton (1996), the environment of collective ecological management is a 'reciprocating' one rather than a 'giving' one, in the sense that

human transformation and intervention in the environment are core aspects of social–environmental interaction. According to Milton, a key part of regarding the environment as a reciprocating one is a cultural recognition of the dependence of human society on the environment.

13 It may be recalled in Chapter 3 that an argument was made connecting defending biodiversity with preserving cultural diversity within the context of the developing world. Here, within the context of the developed world, we have a similiar argument connecting the preservation of a particular sub-climax ecological state with the preservation of a particular cultural self-understanding.

14 In this respect then tending one's garden may be just as effective in demonstrating and developing the 'ecological virtues' as 'wilderness experience'. 'Authentic' care of the non-human world does not reside solely in wilderness protection, or the protection of endangered species.

15 This idea has been developed by the Wuppertal Institute as a basis for calculating sustainability for different regions and countries (Friends of the Earth, 1995). The basic premise of their calculation is that each region and country be allowed that proportion of global carrying capacity (or 'environmental space') which corresponds to its proportion of the global population.

16 A more radical view is given by O'Neill who claims to 'find it difficult to see how environmental goods could be realised without planning within the economic sphere itself. The question is not one of market or plan, but rather what forms of planning are compatible with other goods, in particular, that of autonomy' (1993: 175). O'Neill's point is that a democratically planned economy is not necessarily a prelude to totalitarianism.

17 A similar conclusion was reached by Jefferson who stated that, 'I hold it that a little rebellion, now and then, is a good thing, and as necessary in the political world as storms in the physical . . . It is a medicine necessary for the sound health of government.'

18 Given that the search for a collective strategy/process for managing the commons involves moral, scientific and cultural as well as more obvious political and economic inputs, there is a sense in which this process requires that *all* forms of valuation and understandings of the environment and social relations to it are given 'voice'. Hence science is not privileged relative to indigenous forms of knowledge and traditional commons-type, i.e. non-state, ecological management, which still exist in many parts of the world (Shiva, 1988; Goldsmith et al., 1992).

19 Much of the debate around ecological restoration centres on the practical and philosophical implications of attempting to 'fake nature' (Elliot, 1982). For example, Goodin (1992) holds that a restored landscape is less 'valuable' (in terms of his green theory of value) than a 'natural' one, since the former is analogous to a copy of the latter. Goodin's green theory of value holds that what is valuable about natural entities is their non-human origin; once humans restore landscapes their value is diminished relative to the 'original'. The reason for this value differential between original and restored nature is that the basis of nature's intrinsic value, namely its 'naturalness', is present in the former and not in the latter (Elliot, 1994: 40–3). However, while there is philosophical debate concerning ecological restoration, there is no suggestion that it ought not to be done. Doubt as to whether it ought to be done arises where claims of future restoration of ecosystems or landscapes are used to

justify present ecological disruption: see Goodin (1992: 31–41). However, if we seek to realize values other than the intrinsic value of nature based on its 'naturalness', there may be less objection to the diminished value of restored ecosystems/landscapes relative to the original. For example, if what we wish is to maximize biodiversity then it is clear that a restored ecosystem which had a greater number and diversity of species than before human interference would be more valuable than the original.

20 Extensive ecological restoration will be necessary if global sustainability is to be achieved, in the sense that biodiversity will have to be increased. For example, according to a Friends of the Earth Europe (1995) report, if the European Union is to achieve sustainability there needs to be a tenfold increase in protected land so as to support the levels of biodiversity required.

6

Green Political Economy

CONTENTS

This chapter outlines a green political economy to underwrite the view of green politics as collective ecological management outlined in the previous chapter. The importance of developing a green political economy lies in the fact that the most crucial aspect of social–environmental interaction is the economy–ecology metabolism. That is, given the overwhelming centrality of this metabolism in determining the character and overall ecological impact of any social–environmental relation, analysing present and possible political economic theory and practice is a key part of rethinking green politics.

Before moving on to introduce an account of green political economy which concurs with collective ecological management, I wish to discuss and criticize the two prominent political economic theories used to address ecological issues: neoclassical environmental economics and free market environmentalism.

A Critique of Neoclassical Environmental Economics

Although often equated, neoclassical environmental economics and free market environmentalism represent quite distinct political economic perspectives, assumptions and aims. Despite sharing a broad concern with finding 'economic' and 'market-based' solutions to environmental problems, these two theories of political economy lead to radically different forms of economic organization both of which, from a green point of view, are problematic.

For Mulberg (1992), neoclassical economics' methodological commitment to positive science, although ostensibly in favour of the 'free market', actually constitutes a justification for economic planning rather than *laissez-faire* economics. According to him, 'the logic of positive economics leads to a notion of economic planning . . . In fact, the solutions based upon orthodox theory apply a sort of planning "supplement" to a market analysis' (1992: 335). Within neoclassical environmental economics, this planning supplement is arrived at by using various non-market techniques, such as cost–benefit analysis and contingent valuation exercises, to ascertain individual economic valuations of environmental resources, that is, their economic benefits and costs to the individual. Economic values are arrived at via the creation of *hypothetical* markets in which individuals are asked to trade non-tradable environmental goods. For example, a neoclassical environmental economics approach to dealing with a proposal to develop a particular environment would involve conducting survey research, in order to determine individual 'willingness to pay' to prevent the proposed development or to ascertain the monetary amount they would require as compensation. In other words the neoclassical approach is to conduct a cost–benefit analysis, aggregate individual preferences as expressed in the exercise and make recommendations as to the 'adjustment' of the price mechanism depending on the balance of 'benefits' and 'costs'. The neoclassical approach attempts to ascertain the price of unpriced and non-traded environmental goods and services (Pearce et al., 1989; 1993). Its starting point is that many environmental problems are due to the unpriced character of environmental goods and services, such as clean air, drinkable water and unspoilt countryside. To combat the misperception of environmental public goods being 'free' it proposes to put a price on them so as to internalize a potential externality. In the language of economics, the lack of a market price for such goods means that market behaviour will lead to a suboptimal, or Pareto-inefficient, allocation of these particular resources.[1] In plain words, because these environmental resources do not presently have a price, the operation of the market will tend to overuse and over-exploit them, resulting in excess pollution and other forms of environmental degradation. By giving the environment a price tag, markets will ensure that the many resources and services it provides to the economy and society in general will be 'protected' and sustainably exploited. Ascribing a price to environmental goods and services makes them 'visible' within the market and nature's contribution to economic activity is thereby acknowledged.

For free market environmentalism, on the other hand, environmental problems are largely due to open access. Hence they propose private property rights as a solution to over-exploitation and degradation of the 'ecological commons'. Thus neoclassical environmental economics and free market environmentalism can be distinguished by the different aspect of the 'public goods' problem of the environment each stresses. Free market environmentalism stresses the non-excludable character of environmental public goods, while neoclassical environmental economics emphasizes their unpriced, 'free' character.

Unlike the free market environmentalist solution, environmental economics does not hold that the value of a given environmental resource or amenity can only be revealed in actual market exchange. This fundamental difference can be seen in the distinction drawn by Pearce et al. between the privatization of natural resources and the creation of new 'environmental markets' (the free market environmentalist approach) and the 'modification of markets by *centrally deciding* the value of the environmental services and ensuring that those values are incorporated into the prices of goods and services' (1989: 155, emphasis added). The state or regulatory body using the survey tools of neoclassical environmental economics ascertains the 'social cost' of pollution, for example, and from there the 'economic price' of pollution can be determined. Using this information, the market can be altered by the state to ensure that environmental externalities such as pollution are 'internalized' by market actors.

For free market environmentalism, following the Austrian school of economic theory, the creation of actual rather than hypothetical markets is the only way in which the 'real' or 'true' economic value of an environmental resource can be revealed. They therefore argue for privatizing environmental resources and allowing market exchange to determine the level and organization of the material interaction between economy and environment. Neoclassical environmental economics on the other hand relies on survey research and other techniques such as contingent valuation exercises to cost people's environmental preferences within a hypothetical market situation. From a free market environmentalist perspective 'Opinion polls and surveys are used as proxies, but they do not provide reliable information about the value of environmental amenities because individuals are not faced with actual trade-offs. The respondents bear no actual costs' (Anderson and Leal, 1991: 92). The libertarian economic view is that aggregate data are useless, since all costs and values are known only at the individual level in actual exchange. Another way in which we can distinguish the free

TABLE 6.1 *Three models of political economy*

Value formation	***Theory***	***Institutions***	***Causes of ecological problems***
Subjective	Free market environmentalism	Free market	Policy failure, 'tragedy of the commons'
	Libertarianism, Austrian school	Organic institutions	Lack of clear, enforceable and tradable private property rights
Objective	Neoclassical environmental economics Green social democracy Ecological modernization	State planning supplement	Market failure
Intersubjective	Green political economy	Collective ecological management	Hegemony of economic interests Environmental preferences viewed as private and given

market environmentalism and environmental economics approaches is to say that the former seeks the creation of actual markets for private and privatized environmental goods, while the latter creates proxy markets for public environmental goods.[2] Free market environmentalism decomposes the public goods dimension of environmental problems into discrete private goods problems which can be 'solved' either on the open market or through private litigation. As will be argued below, it depoliticizes environmental problems as a consequence of its distrust of the state and its unbounded faith in the environmental virtues of the free market.

Neoclassical environmental economics has been criticized on the grounds that the separation of the economy into planned and non-planned sectors is arbitrary (Mulberg, 1992). If the economic value of environmental resources can be determined in hypothetical markets, why not extend this process to other sectors of the economy? If there is no need for the market to determine the appropriate price structure in respect of environmental resources, why not abolish the market mechanism and introduce planning on the basis of objective economic valuations in other areas? The objectivity of the neoclassical 'hypothetical market' approach lies in the fact that the initial generation of price or monetary value is centrally decided, which is then used in exercises such as cost–benefit analysis and contingent valuation. The aggregate of these *individual valuations* is then assumed to approximate the *social or collective valuation* of these environmental goods and bads (see Table 6.1). Yet, as suggested in earlier chapters, the public goods character of the environment implies that what public policy-making requires is judgements about the public good, rather than private calculation about it as if it were a private good.

These objective valuations are used as the basis of environmental policy-making by the state or its agencies. The institutional theory underpinning the neoclassical model is one which leaves the state and its agencies in the pre-eminent position to both determine the economy–ecology metabolism and implement the policies required to achieve that chosen rate of metabolic exchange. The argument here is not against the attempt to politically regulate environmental resource use, but rather is against the particular way in which neoclassical environmental economics conceives of environmental management, and picks out the appropriate decision-making institution. In short, the institutional dimension of environmental economics is very bureaucratic and overly state-centred with little meaningful input by citizens in the decision-making process. Its state-centred and planning aspects thus make it the political economy of ecological modernization, as suggested in the last chapter.[3] It is the state and its experts, rather than citizens, which determine the content of environmental management, through the manipulation of the price mechanism. It is important to point out that within this model the question is about 'fine-tuning' the market; the options available do not include questioning the market itself as the most appropriate institution for managing environmental resources.

In contingent valuation experiments people are asked as private consumers rather than citizens to reveal the private cost/benefit to them of particular environments or species or some environmental amenity (Sagoff, 1988; O'Neill, 1993; Keat, 1994; Jacobs, 1997). The assumption here, as Jacobs (1997) has pointed out, is that individuals' preferences are assumed to be exogenous. The aim of contingent valuation is then to 'reveal' these already formed and given preferences. That preferences for environmental goods are not given, but require an intersubjective context within which individuals attempt to value environmental resources as *public goods*, is ruled out on *a priori* methodological grounds. That is, neoclassical environmental economics misrepresents how people think about environmental issues by using an economic methodology developed for assessing individual preferences for privately produced and consumed goods and services and applying it to a set of issues to which it is inappropriate (Jacobs, 1997). Contingent valuation gives us the wrong information upon which to make environmental decisions. According to Jacobs, contingent valuation 'asks the wrong question. Asking the personal question "how much are you willing to pay?" encourages people into a self-interested stance. This is the appropriate question in a market for a private good' (1997: 217). But since the environment and its services are not produced by anyone and are public goods, their social valuation demands a process of citizen deliberation rather than the state and its experts using consumer preferences to determine environmental preferences (i.e. prices). The neoclassical model of political economy uses an objective rather than an intersubjective method to 'discover' the 'social valuation' of environmental resources. Consequently the relationship between economy and environment is determined *for* individuals *qua* consumers (by the state manipulating the price mechanism on the basis of information regarding consumer preferences) rather than *by* individuals themselves collectively *qua* citizens.

Attitudes to public goods can only be ascertained through public fora rather than through market or quasi-market techniques. For it is only within fora that individuals can actually be said to formulate their judgements about the public good. As private consumers they only need to consider *their* own calculation of the costs and benefits to *themselves*. However, with regard to public goods, decision-making requires a setting within which a *variety of possible* calculations concerning the possible effects (costs and benefits) on the *collective* can be considered. In short, decisions about issues such as pollution prevention (which is a public good) require political not market institutional settings for public policy-making. Environmental public policy decisions ought to be based on asking people as *citizens* for their *judgements* concerning the public good, and not as consumers interested only in their own good. As citizens, individuals are not asked to take a principled non-self-interested stance; rather they are asked to ascertain a wider sense of self-interest than that appropriate as consumers. The point is that it is not appropriate to ask people to think and act as private consumers when dealing with environmental public goods.

Self-interest *qua* consumer is not the same as self-interest *qua* citizen. The more one's decisions affect others (fellow citizens, non-humans, future generations, non-citizens) the less appropriate is a market or economic approach to deciding the fundamental nature of environmental problems. While many environmental problems are also economic ones, this does not exhaust the scope of human interests that pertain to the issue. Since environmental decision-making does affect the interests of those who are often not a party to the decision-making process, it is reasonable that individuals be asked to consider the likely effects of environmental change on the interests of others.

At the same time there is the argument that it is not appropriate to reduce normative claims or valuations to monetary values. Here market or quasi-market approaches to environmental public policy-making can be criticized on the grounds of the incommensurability of valuations one can make in respect to the environment. As Keat (1994) notes, it is simply inappropriate to reduce moral valuations to monetary sums. Treating judgements about what is right as individual preferences is inappropriate (1994: 338). That is, one cannot reduce moral judgements to economic criteria. Asking someone to put a monetary amount on their friendships or their relationship with their dog, or the cash value of a landscape, is to seriously misrepresent these relationships. Evidence suggests that people do not find it easy or desirable to express their views about environmental issues in monetary terms. The widespread phenomenon of 'protest bids' within contingent valuation exercises, where individuals either refuse to put a monetary amount on an ecosystem for example, or place an astronomically high monetary value on it, can be taken as evidence of 'ethical commitment' rather than economic calculation (O'Neill, 1993; Vadnjal and O'Connor, 1994; Splash and Hanley, 1995). Protest bids indicate that people refuse to view the environment as a commodity. Yet one of the consequences of neoclassical environmental economics is to commodify the environment, so that its 'economic value' may be discovered and used by the public authority to correct market imperfections (Pearce et al., 1989; 1993). A key claim of neoclassical environmental economics is that treating the environment as 'natural capital' provides the strongest case for environmental protection.

What can be deduced from protest bids is that to assume the commensurability of economic preferences and moral judgements is to commit a serious category mistake. The normative dimension of social–environmental issues can be illustrated by comparing the argument here with that in Chapter 3 where I suggested that the use of 'rights' within environmental discourse can be understood as a way in which the moral intensity with which some individuals feel about the treatment of the non-human world can be communicated. This communicative dimension is crucial. Now whether or not one agrees with those who argue for the rights of animals, ecosystems etc., one can at least be sure that what is being communicated is a *moral* argument. One may not agree with those who use this particular moral idiom to argue for certain treatment of the non-human world, but one is at least sure that what is at issue is a moral question. In the case of treating moral judgements as economic preferences, a category mistake is committed. The criticism of aggregative/consumer approaches to environmental public policy is simply that they are not the appropriate idiom and do not provide the appropriate information for dealing with these issues. Normative questions demand normative contexts. To use the consumer/citizen distinction, the point is *not* that consumer preferences are rejected as inappropriate and substituted by citizen judgements when making environmental decisions. This would, as Keat (1994) points out, neglect to see that consumer preferences themselves have evaluative significance, and that the market/economistic approach to environmental valuation is not, as neoclassical economics itself holds, 'value-neutral' or 'positive'. Criticizing those who simply reject a 'consumer' approach to environmental issues in favour of a 'citizen' one, Keat holds:

> The impression is thereby created that the economistic approach does *not* express or rely upon any such ethical or evaluative principles. Yet this is misleading. For, whatever its defects, it seems clear that there is *some* such theory involved – a broadly utilitarian one according to which the right action is that which maximises aggregative human well-being, where the latter is itself understood as consisting in the satisfaction of preferences. (1994: 340)

The point is that the normative underpinnings of the neoclassical perspective are what are to be contrasted with alternative normative perspectives. This is a reason why it was suggested earlier that an adequate understanding of the normative dimensions of social–environmental affairs *does not* require the rejection of economic valuations of the environment. What is required is the *translation* of economic valuations (preferences) into normative claims. Unearthing the normative underpinnings of preferences was, it will be recalled, the central aim of the ethics of use. There it was argued that only by finding the values which ground preferences normatively can the distinction between 'serious' and 'non-serious' human interests be made, and legitimate human use of the environment be distinguished from illegitimate abuse. Preferences within the context of social–environmental affairs are necessary but insufficient to fully justify particular human uses of the environment. One way of looking at this has been suggested by Norton:

> preference models provide only one approach to the valuation problem and the usefulness of preference explanations is actually enhanced if they are regarded as describing only one aspect of environmental valuation. When the study of preferences is supplemented with a broader, more comprehensive treatment of other aspects of environmental values, the overall picture of environmental valuation is clarified. (1994: 314)

Rather than seeking a single commensurable unit (money values) upon which to base environmental decision-making, we ought to search for a common framework within which all valuations can be articulated. That is allow the plurality of environmental values to be articulated, and then assessing them via a public judgement of their normative underpinnings, perhaps through voting on them. The information generated from this exercise can then be used to make decisions, or at the very least used to 'temper' or act as a 'side-constraint' on 'pure' monetary valuations of environmental goods. Rather than make decisions on the basis of counting and comparing monetized environmental preferences, publicly comparing options directly in terms of their desirability is often more appropriate. A shorthand version of this for environmental decision-making might therefore be: 'Think of the issue in terms of right and wrong first, and then in terms of costs and benefits.' Note that this does *not* mean that we ought to think of environmental decision-making *only* in moral terms.

The point is that one needs to be sensitive to the processes and contexts within which preferences are formed and re-formed; accepting them as 'given' and/or 'fixed' is often inappropriate. Since people behave and value differently under different institutional contexts, preferences are not exogenous but, in part, formed under specific institutional contexts. The root of the economic, market-based approach is that it assumes we ought not to criticize individual preferences, since to do so would be to violate its professed value neutrality. It is the positivism of neoclassical economic theory which needs to be criticized as much as its assumption of the commensurability of values. Reducing values to prices is to distort those values. Whereas according to Wilde the cynic 'knows the price of everything and the value of nothing', the economist knows the 'value' of everything because she reduces 'value' to the common denominator of money.

Free Market Environmentalism

In contrast to the neoclassicist, the free market environmentalist approaches environmental problems as stemming from a lack of clear, enforceable and tradable property rights. Typically, environmental problems are held to result from the 'tragedy of the commons', the overuse of a resource which no-one owns (the seas) or everyone owns (state-regulated resources). It is important to point out that the 'tragedy of the commons', as famously expressed by Hardin (1968), actually represents an open-access system *not* a commons regime. As many ecologists have pointed out, commons regimes do regulate access to and use of commons resources, but do so without recourse to either the state or the market (Goldsmith et al., 1992; Wall, 1994). Thus, commons regimes do not necessarily lead to resource over-exploitation such that privatizing them is the only or most appropriate solution. The argument that only the creation of a market in environmental goods (and bads) will ensure a socially optimal level of environmental protection and sustainability thus begins from a debatable conceptualization of the problem. As mentioned earlier, a problem with the free market environmentalist approach is that it regards environmental problems as allocative ones relating to *private* environmental goods. While market logic may be appropriate in deciding issues concerning private goods and services, it is not appropriate with regard to public goods, such as environmental quality and protection.

In terms of the social valuation of environmental resources, the market-based approach is the standard one that market exchange will reveal the 'true' value of the environment. By aggregation of individual preferences for environmental resources, as 'revealed' by supply and demand intersecting at the equilibrium price, the 'efficient' economy–ecology metabolism will be determined. Unfortunately, an economically 'efficient' metabolism may not be an ecologically sustainable one (Daly, 1987). The problem with the free market environmentalist approach is that, in common with the neoclassical approach, it misrepresents the issue by reducing the question of economy–ecology interaction to a matter of the market-efficient allocation of environmental resources.

Values are reduced to preferences and aggregated to approximate the social valuation of environmental resources (understood in economic terms). That preferences for environmental resources may not be 'given' to be 'revealed' in market exchange, but may rather be created and subject to how others value the environment, is not considered. However, as indicated above, there is evidence that individuals do not subjectively value the environment in terms of private costs and benefits to them, but rather perceive environmental resources as public goods and express this in moral not economic terms.

In terms of Table 6.1 free market environmentalism is premised on a subjective account of value. It is the aggregative activity of private individuals *qua* consumers and producers which determines the economy–ecology metabolism. Pure economic rationality is assumed to lead to an ecologically rational outcome. For many environmental public goods, for which no market can exist, such as biodiversity protection and prevention of global warming, it is clear that a free market environmentalist approach is fundamentally flawed. That is, there are environmental problems whose nature is such that they cannot be disaggregated into component parts which can be solved by exchange on the open market between property holders. Such a reductionist, non-integrated approach to environmental problems is, as Dryzek (1987) points out, unable to adequately deal with the integrated, interdependent character of social–ecological problems.

What is missing from the free market environmentalist perspective is the intersubjective nature of environmental valuation, based on the idea that the environment is seen as a public goods question and not a matter for private, economic calculation. It is important to point out that not only the institution (the market) but also the language and type of information used within free market environmentalism can be questioned. As pointed out in respect of ecological modernization, for free market environmentalism, economic valuations are the only admissible forms of valuation, and prices in the market the only form of signalling used to manage environmental resources. In reducing values to preferences, free market environmentalism also 'crowds out' non-economic forms of environmental valuation, such as political, cultural or moral considerations which require intersubjective and deliberative rather than subjective and aggregative institutions. In dealing with social–environmental interaction, the 'social' cannot be reduced to the 'economic', or the 'environmental' to the category of 'economic resource' or 'commodity'.

Recalling the argument in Chapter 3, free market environmentalism also displaces the political-normative question of determining 'use' from 'abuse'. It does this by simply assuming that an environmental good or service is a privately owned resource without reference to whether it may be a 'proscribed' resource or public resource. An example of the former is the attempt by the biotechnology industry to establish exclusive intellectual property rights over genetic material, within the international debate concerning biodiversity. A clear example of the latter, where environmental goods are a public resource, is the 'Wise Use' movement in America which has lobbied Congress to privatize federal land.[4] This would transform these environmental goods from public goods in which their 'use' was limited to access and recreation, into private goods where use would be much more extensive, intrusive and ecologically damaging. 'Use' for the Wise Use movement implies forms of economic development, such as mining, logging, building, hunting and generally 'unrestricted access to all natural resources for [private] economic use, benefit and profit' (Callahan, 1992: 2). From this free market environmentalist point of view, objections to such 'abuses' can only take the form of other private agents, such as environmental protection organizations, purchasing these lands themselves and thus preserving them. This has the effect of making environmental protection a function of wealth, where only those with sufficient purchasing power can buy it. For Eckersley, the fact that free market environmentalists are not interested in the distribution of private environmental property rights:

> must be seen as a thinly disguised endorsement of the existing distribution of property rights and income . . . Indeed, the long-term consequence of zealously pursuing the privatisation of environmental resources . . . is likely to be the intensification of the already wide gap between the propertied and the propertyless, and the rich and the poor, both within and between nations. (1993b: 15)

However, just as there are positive aspects of the orthodox neoclassical solution, namely the idea of environmental regulation, likewise there may be some positive aspects that may be taken from free market environmentalism. One is the notion of stewardship and care implicit in the emphasis on private property and ownership. As Anderson and Leal (1991: 3) point out, the idea of having property, a claim of ownership, to the land or some environmental resource involves a commitment to look after it. However, as pointed out above with reference to the 'tragedy of the commons', the stewardship ideal implicit in the idea of property is not confined to *private* ownership. *Common* ownership can also deliver the virtues of stewardship and careful husbanding. As Goldsmith et al. (1992) argue, there is plenty of empirical evidence from around the world that commons regimes can deliver sustainable levels of resource exploitation. As has been suggested the 'tragedy of the commons' is as a result of 'open access' to an environmental resource; a commons regime is not a 'free-for-all' (Goldsmith et al., 1992: 127). Indeed, according to this line of argument, so-called environmental tragedies of the commons are in fact the result of tragedies of enclosure (Bromley, 1991). It is only if one accepts the initial argument that environmental problems are caused by common ownership that private property in environmental resources can be regarded as the only solution. Commons regimes in which the commons are the property of all, or all members have an equal right of access, are a different form of collective ownership than state ownership, which is the real target of free market environmentalism. Under central state ownership, it is less likely that individuals will feel they, conjointly with others, actually own the resource and thus have responsibility for it. Under these conditions, it is truer to say that the resource is owned by nobody. *Collective* ownership and regulation of environmental resources (which is not the same as centralized state ownership and control) do not stand condemned on ecological grounds.[5] For example, one could envisage a hybrid 'commons-type regime', where environmental resources were managed by a combination of local state co-ordination together with local community, including business, participation.

A real-world approximation to this is the Local Agenda 21 initiative, which on one level can be seen as an attempt to include the local citizens as 'stakeholders' in the regulation and management of local social–environmental affairs. Local Agenda 21, which was signed at the Rio Earth Summit (UNCED, 1992), proposed that local authority activity is central in achieving sustainable development. In short, LA21 requires local authorities to consult with the various 'stakeholders' within their jurisdiction – industry, trade unions, non-governmental organizations, the local community – and by 1996 come up with an environmental agenda for the local area for the next century (Gordon, 1993). Although still very much in its infancy, Local Agenda 21 does exhibit aspects of what many regard as key issues central to the achievement of sustainability as a policy goal. These issues include community environmental education, democratization, balanced partnerships between public and private sectors, and integrated policy-making (Agyeman and Evans, 1994: 20–2). Other aspects of Agenda 21 include the role of local government in mobilizing citizens, devolved land-use planning decisions as well as increasing the role of local government in national and international environmental policy-making (Gordon, 1993: 152).

Stated in these terms, Local Agenda 21 is close to some of the themes and aims of collective ecological management outlined in Chapter 5. The emphasis on bottom-up, participatory forms of decision-making, together with environmental education, does suggest that Local Agenda 21 can function as one institutional facet of a wider process of democratic ecological governance. The emphasis in Agenda 21 on identifying 'stakeholders', which can be expanded to a broader concern with 'stakeholding' in which non-humans and future generations can be considered stakeholders (Roddick and Dodds, 1993), highlights the role of local citizens together with the local authority as stewards of the local environment, and the contribution they can make to global ecological sustainability. Coupling such local policy responses to environmental problems with local forms of economic activity such as LETS (see later), for example, is suggestive of a modern 'commons-type' regime which can secure an ecologically rational form of collective ecological management. Making people aware of the interconnectedness of human well-being (including economic considerations) and the environment, while also giving them a greater say in formulating local environmental policy, does highlight the connection between long-term human self-interest and environmental responsibility, which is a central aspect of ecological stewardship. Local Agenda 21, in conclusion, can encourage ecological stewardship since the virtues of stewardship need not be tied to ownership of, and direct productive relations with, the environment. Having a stake (*qua* citizen) in *managing* social–environmental relations is a more realistic way of thinking about how to create a sense of ecological stewardship than seeing the issue in terms of actual *ownership* of the environment (*qua* property holder).[6] Here the emphasis within the process of Local Agenda 21 on 'empowering' citizens (Young, 1993: 109) is as crucial as educating them.

One type of private ownership of environmental resources which might be seen as positive from a green point of view is the private ownership of the land as part of farming viewed as a social practice. According to Thompson:

> Stewardship does not arise as a constraint on the farmer's ownership and dominion of the land, but as a character trait, a virtue, that all farmers would hope to realise in service to the self-interests created by ownership of the land. (1995: 74)[7]

In this instance the argument for private ownership of the land is not for the same reasons as put forward by free market environmentalism. In the case of agricultural stewardship, private ownership is not justified on the grounds of economic productivity or private profit alone, nor is its content determined by market exchange. Indeed, according to Thompson (1995), the family-owned farm, properly speaking, restrains the productivist imperative which would transform agri*culture* into 'agri-business'. That is, private ownership (or secure tenure) of the land, within the context of farming as a way of life and not simply as an industry, may be justified from a green point of view. However, once the 'biocultural' context within which farming was a social practice of land stewardship becomes eroded, the issue of land ownership cannot be answered without raising the question of the ethical status of the transformative use of the land in terms of 'use' and 'abuse'. Private ownership within the context of a social practice is not the same as private ownership within the context of a market system. As pointed out in the last chapter, collective ecological management is understood as a cultural and moral as much as an economic or political process concerning social–environmental interaction. Hence the green support for organic farming is not simply for the use of a particular ecologically sensitive technique of agricultural husbandry, but at root a call for a return to farming as a social practice, a way of life which expresses particular biocultural values and virtues, and 'traditional' forms of farming knowledge and techniques, including, most importantly, less ecologically damaging forms of farming and animal husbandry.

One way of looking at this is that any decision made on the economic use of the land must be derived from, rather than independent from, a prior settling of the political-normative issue concerning social–environmental interaction. In other words, from a green point of view, ownership relations cannot be based or justified on economic considerations alone. Within the context of collective ecological management, they must answer to political and moral considerations. There is nothing particularly novel in this. If property rights and markets themselves are politically created and maintained, they can, and from a green perspective ought to, be politically and morally constrained. Market-based solutions to social–environmental problems, such as that proposed by free market environmentalism, do not represent 'depoliticized' solutions. As contemporary political debates around environmental problems demonstrate, and as suggested in the last section, market-based approaches are as inherently political and just as normative as any of the alternatives.[8] The fundamental debate is thus at the level of the moral values and political principles of free market environmentalism and its alternatives, environmental economics and collective ecological management, rather than at the level of institutional structures alone.

Having raised objections to both neoclassical environmental economics and free market environmentalism, in the next section I outline an alternative 'green political economy'.

Green Political Economy

According to Mulberg in his critique of neoclassical environmental economics, 'Given that the non-traded environmental resources are not capable of objective valuation, what becomes important is the *process* whereby the subjective valuations are made known' (1992: 340). While agreeing with this, I wish to also suggest that environmental valuation is a matter of choosing institutions within which preference formulation is central, rather than aggregating given preferences. The social valuation of environmental resources (including such basic questions as what is to count as a 'resource') is thus, at root, a matter of choosing the institutional setting within which such valuations are formed.

Recalling the earlier discussion of how being human carries with it broad ways of viewing and relating to the world, unlike traditional theories of political economy, green political economy is concerned with the normative question of what is to count as an economic resource. 'Resources' as much as 'preferences' within green political economy are not taken as exogenously given. This is a different, and more radical, sense of resources not being 'given', i.e. infinite in quantity and always available, which characterized the early 'limits to growth' green movement. Substituting 'resources' for 'food', Illich's statement that 'It is human to see the environment made up of three kinds of things: foods, proscribed edibles and non-food' (1981: 29), expresses this normative dimension of green political economy. In this way one can view the radical deep ecology argument for 'wilderness' as a call for transforming current resources into 'proscribed resources'. Another example is how technological and scientific development transforms what was previously a 'non-resource' into a possible resource. From a green political economy point of view, a collective judgement has to be made as to whether it is to count as a 'resource' or a 'proscribed resource'. An obvious example of this is biotechnology, where genetic information is now poised between those who see it as a legitimate new resource, and those who raise normative objections to this development, and wish to either permit the technology to develop but place genetic information in the category of 'proscribed resource', or abandon the technology completely and maintain genetic information as a 'non-resource'.[9] A more historical example is the movement of human slave labour from the category of 'resource' to that of 'proscribed resource'. The same argument is used by contemporary animal rights activists who in describing animals as 'slaves' not only seek to convey their moral condemnation of this situation, but do so by tapping into a well-established and intuitively appealing moral tradition and idiom.

Unlike both neoclassical environmental economics and free market environmentalism, green political economy is distinguished by holding that what environmental management requires, initially, is a political and deliberative process by which collective valuations of the environment can be articulated, and used as the basis of determining the economy–ecology metabolism. This is an intersubjective process of deliberation. What it ultimately comes down to is that institutional settings are required in which individuals are asked to make *judgements* about how *we* as a collective are to value/use the environment, as opposed to simply expressing 'willingness to pay' (neoclassical environmental economics) or private economic calculations of environmental resources (free market environmentalism). This is another way of saying that the rationality that characterizes green political economy is, as indicated in the last chapter, ecological rationality. The rationality of green political economy in seeking to establish sustainable and symbiotic social–environmental relations can be understood as a form of communicative rationality within which instrumental rationality is nested. That is, a pure instrumental orientation towards the non-human world is not rejected within green political economy. Rather it seeks to 'recalibrate' this orientation according to the (intrahuman) communicative norms of an ethics of use. This, as argued later, is what is meant by green political economy seeking to 're-embed' the economy within society, as a necessary prelude to reintegrating economy and ecology.

Arguing for the priority of an intersubjective approach to environmental valuation does not mean that this institutional structure will be applied over the full range of economy–ecology affairs. In other words, the wholesale politicization of the economy is not envisaged. On the one hand, the politicization of economy–ecology interaction is not, *pace* free market environmentalism, equated with state administration. Rather there are a range of possible institutional designs which would democratize environmental management without meaning that each citizen has to vote or actively participate in determining every aspect of the economy–ecology metabolism. The idea that only direct democracy is consistent with green politics will be critically examined in the next chapter where a distinction is drawn between deliberative, direct and participatory forms of democracy. The real issues are democratic accountability and opportunities for citizen participation and input into decision-making processes pertaining to economy–ecology interactions, in ways by which the economy–ecology metabolism is not solely 'managed' by a *self-regulating* market system.

The institutional setting for determining the metabolism ought to be appropriate to the particular issue at hand. As green theorists and political economists have pointed out, there needs to be a distinction drawn between macro- and micro-level economic activity (Porritt, 1984; Daly and Cobb, 1990) and, as pointed out in the last chapter, between decision-making and decision-recommending institutions.[10] For example, macro-economic issues such as threshold levels for the environmental impact of the economy, and extra-economic, political-normative ones as to *what* is to count as a resource and *how* resources are to be used, can be determined politically, not so much *by* the state as *through* political institutions. Clearly, the institution should be appropriate to the type of social–environmental issue under consideration, as suggested by the implementation of Local Agenda 21s. Collective ecological management and green political economy do not suppose that the nation-state and its agencies are the most appropriate institutions for dealing with all environmental problems. As Lindblom has pointed out, centralized institutions have 'strong thumbs, no fingers' (1977: 76–89), but as Dryzek suggests, ecological rationality demands 'nimble green fingers' (1987: 109) as much as regulatory powers. Institutional examples of such nimble fingers include local environmental management strategies, citizens' juries, and public enquiries into land use which set their own agenda, with powers for example to set the parameters within which environmental management is to take place. That is, these deliberative, consultative bodies do not themselves make decisions concerning the actual details of environmental management, but are decision-recommending bodies, made up of various groups of 'stakeholders' (Jacobs, 1996). At the same time, where there are already existing commons regimes or the possibility of creating one, these regimes would also figure as another nimble and green finger to complement the thumbs of central government. Indeed, as suggested earlier, it may be possible to integrate local state environmental regulation within a commons-type regime as a form of ecological governance.

Actual economic organization and micro-level decision-making may be left to the market (Jacobs, 1991) or to non-market institutions in civil society (O'Neill, 1993; Achterberg, 1996), including communities (Goldsmith et al., 1992; Fairlie et al., 1995), as well as the regulatory and legislative processes of local- and national-level state institutions. Once the political-normative task of deciding which institutional processes are appropriate to which aspects of managing the economy–ecology metabolism, various institutional settings and principles at different levels may be used. In conjunction with the distinction between macro- and micro-levels it may be that an ecologically rational economy–ecology metabolism calls for the division of the economy along functional and ecological lines as a precondition for effective environmental management. In terms of the institutions for the governance of the economy–ecology metabolism, it may be expected that collective ecological management will make use of market, state and substate institutions and those associated with community, as well as combinations of them. That is, ecological governance is a multi-institutional political process.

Within intersubjective valuation of environmental resources, the model of the relationship between state and civil society is more appropriate than that between state and market. In other words, the institutional focus of collective ecological management and green political economy lies in the relationship between state (local and national), non-state (market) as well as non-market civil institutions (O'Neill, 1993) and social practices. Green political economy, unlike orthodox economic theory, is an institutional theory of economics (Jacobs, 1994; Dryzek, 1996). That is, green political economy rejects the methodological individualism which underpins the neoclassical and free market environmentalist approaches, seeing, as Jacobs suggests, 'that economic behaviour is culturally determined, and that institutions in society (such as governments, regulations and property rights) are not "market imperfections" but the very structures which allow markets to operate' (1994: 84). The Local Agenda 21 process discussed above, and ecological management and regulation by central and local state institutions, discussed earlier, are examples of this institutional focus of green political economy. This institutional dimension also relates to the critique above of the neoclassical and free market environmentalism views of environmental valuation. Both assume the market as the appropriate institutional setting, one effect of which is the reduction of values to preferences. In part, this institutional focus marks green political economy as returning to the tradition of classical political economy. Other classical political economy themes include its focus on moral virtue, the question of the relationship between economic organization and the social order expanded to include the social–environmental order, as well as explicit attention to the political-normative context of economy–ecology relations. Part of what this involves is a reconceptualization of the sphere of the 'political', expanding it beyond a state-centric view (Bookchin, 1992a), as well as a reconceptualization not only of the 'economy' as suggested by Mellor (1992; 1995) and Hayward (1995), but also of the 'market' (Polanyi, 1957).

The 'Market', Capitalism and Markets

Following Polanyi, we may say that there is no such thing as *the* market, rather there is a 'market system', by which he meant a system of *self-regulating markets*. According to him, 'Market economy implies a self-regulating system of markets; in slightly more technical terms, it is an economy directed by market prices and nothing but market prices' (1957: 43). As should be clear by now, green political economy is sceptical of the claim that the material metabolism between economy and environment be regulated primarily by the market system and the price mechanism. With Polanyi and Mulberg, and against the Austrian school of economic theory, green political economy holds that the market system is not an 'organic' or spontaneous creation. Rather as Mulberg points out, 'Markets are simply exchange mechanisms set up by the polity and governed through the legislature. To view the market as "free" or "natural" is reification' (1992: 340).

It is the self-regulating principle of the pure 'free' market system which is problematic from a green political economy point of view, since it removes political and other non-economic considerations from determining economy–ecology relations. This self-regulating, self-referential aspect of the market economy also disembeds the economy from society. In the language of systems theory, the logic of the self-regulating market system is to separate itself from the wider social and political systems to become an autonomous subsystem in its own right (Habermas, 1974; Offe, 1984). The green suspicion of the modern market economy had been eloquently expressed by Polanyi:

> the idea of a self-adjusting market implied a stark utopia. Such an institution could not exist for any length of time without annihilating the human and natural substance of society; it would have physically destroyed man and transformed his surroundings into a wilderness. (1957: 3)

A society in which a system of self-regulating markets was the main social choice mechanism would be not just ecologically irrational, but also socially destructive. Like the world of perfectly competitive markets which are used to justify and explain orthodox economic theory, a completely self-regulating market system is an equally abstract, if more dangerous, fiction. As Gorz notes, such a vision, which he claims is at the heart of capitalism, represents economic rationality finally set free of all restraint (1989: 122).

What I want to argue here is that it is not markets *per se* that greens object to, or ought to object to, but certain features of the contemporary capitalist market system which mitigate against the resolution of the ecological crisis and the realization of green values. A revised understanding of the market, as an uncoerced mechanism of exchange, can find a legitimate place at the heart of green political economy as one institutional setting of collective ecological management. It is the structural features of the globalized capitalist market that constitute the real obstacles to the integration of the economy and the wider economy of nature. On this reading, green politics is anti-capitalist in the sense that the imperative for capital accumulation as expressed in the imperative for economic growth and the operation of the global market economy is incompatible with the green assertion of ecological limits to growth and the importance of socially re-embedding the economy by democratically managing it. It is also anti-capitalist to the extent that it criticizes the way in which values are reduced to prices within the operation of the capitalist market. And it is anti-capitalist in being suspicious of the ecological and political effects of large, multinational corporations in the global economy. But although it may be anti-capitalist, this does not necessarily mean that it is against the institution of the market. The question is rather in what ways this conceptualization of the market as a social institution for uncoerced economic exchange can be a part of collective ecological management. That is, the object is to find ways in which the market can contribute to, or at least not compromise, an ecologically rational metabolism.

Addressing the global dimensions of the contemporary market system, Sachs outlines a good starting point for discussion. According to him:

> we have to finally abandon the idea of a homogeneous unified market from the village to the global level, where the factors of production can be freely moved around, and to conceive of restricted markets, where political

> norms limit the scope and range of market activities without emasculating their potential for innovation and liberty. (1990: 336)

Sachs's argument presents the green case for delinking from the world economy, discussed below, and is largely a reactive or defensive argument for constraining the presently existing globalized market. However, alongside this defensive argument there is also a more positive sense of the market within green political economy. This is a conception of the market which is qualitatively as well as quantitatively different from the present market economy. At the same time this discussion of the market differs from those of Jacobs (1991; 1995), Eckersley (1992a; 1993a) and de Geus (1996), who are close to the ecological modernization model on this issue. It differs by not taking the presently constructed market system as 'given' or the best means by which to realize ecological ends. As will become clear, this is partly to do with the idea that an ecological reconceptualization of the 'economy' means that current conceptions of the 'economy', and thus the market, are incomplete from a green political economy viewpoint. Here I follow a theme within green political economy suggested by writers such as Illich (1981), Hayward (1995) and Mellor (1995), in which the relationship between the 'informal' and 'formal' economies, and the eco-feminist focus on the relationship between production and reproduction, are central to understanding the relationship between economy and ecology.

This 'green' conceptualization of the market draws heavily on local, community-based economic practices and systems that one finds throughout the literature on green political economy. The non-capitalist market within green political economy is generally understood to refer to the operation of voluntary exchange primarily at the level of the local economy. Examples of this market institution include local forms of money systems such as local employment and trading systems (LETS) (Greco, 1994; Lang, 1994; Williams, 1995), community economic development strategies (Shragge, 1993), co-operatives and alternative producer–consumer relations, and combinations of municipal economic and political governance of the local economy (Mellor, 1995). What all these local economic systems share, apart from their shared identity as examples of non-capitalist market institutions, is the decentralist aim of attempting to make local economies as self-reliant and self-determining as possible. This aim, to make local economies as autonomous as possible, is most often expressed within green literature as the idea that local needs should be met locally.[11] In this understanding of the market economy, the primary virtue of the market rests in the voluntary exchange of goods and services it facilitates, and its role in encouraging innovation, rather than as facilitating the process of capital accumulation. A market whose primary purpose is to facilitate trade and exchange at the local level, keeping locally produced wealth locally, as well as meeting needs locally as much as possible, while not completely eroding the potential for capital accumulation at this level, does make a significant break with the economic logic of accumulation and extensive trade which characterizes the contemporary global capitalist system. However, quite apart from the tendency of *unregulated* markets to encourage an economy–ecology metabolism which is unlikely to be 'ecologically rational', in the narrow sense of sustainability, the green stress on local production and limiting trade is also advanced for other reasons to do with green views about autonomy and freedom, as argued below.

The Local Market Economy, Trade and Globalization

Local employment and trading systems (LETS) can be viewed as a non-capitalist market economy, the main characteristic of which is that the medium of exchange is created and regulated at the local rather than the national level. It is geared towards exchange and trade rather than accumulation, and rather than representing an alternative currency system, it is closer to the mark to describe it as an extended or credit barter system. In a LETS a local medium of exchange is created along with a directory of members' offers and wants, which operates as an exchange system matching wants and offers facilitated by the local (nominal) exchange unit. In terms of the distinction drawn earlier between macro- and micro-economic levels, the LETS economy is a market economy at the local level but one where there is no strict boundary between economic and social spheres. It represents the so-called 'informal' economy of everyday life, sometimes called the 'social economy', in which trade and exchange are neither related to nor directly dependent upon the 'formal' market system or the state (Illich, 1981; Henderson et al., 1990; Latouche, 1993).[12] In this section I wish to use LETS as a way to present some of the key aspects of green political economy and its conception of the market. In particular, there is a need to focus on the local as a way of addressing global environmental problems.

The LETS economy is a type of market economy unlike the formal market, in which the benefits of an uncoerced institutional mechanism for trade and exchange can be enjoyed by keeping the scope of the market within socially and geographically defined limits. The economics of LETS is anti-accumulation in the sense that the main purpose of the system is to facilitate trade and exchange within a closed system. Trade is confined within a local network of individuals, since the medium of exchange is only valid within that system. The LETS local market economy is geared to encouraging the circulation of local currency within the local economy, thus stimulating exchange, employment, production and moderate consumption, rather than accumulating wealth in the form of local currency credits (Greco, 1994; Lang, 1994). It is a bounded economy, delimited by membership and place.

Meeting local needs locally avoids the ecological degradation caused by global patterns of trade which require vast transport and infrastructure systems. In such an economy the 'externalities' of pollution and other forms of ecological degradation may be prevented from arising in the first place, because the scale of the economy modifies its environmental impact, while the move towards economic self-reliance also increases the local economy's dependence upon, and impact on, the local environment. While economies of scale would of course be taken into account in deciding economic activities, the point is that these economic issues would not automatically trump non-economic considerations. Rather, 'economies of scale' would have to be judged relative to ecological considerations of environmental impact, long-term sustainability etc., so that what seems an economy (i.e. a benefit) is not in fact an ecological cost. While it makes economic sense (in terms of efficiency and maximization of production) to make economies of scale an important factor in taking economic decisions, it may not make ecological sense, in terms of sustainability, since this involves considerations of *optimality* rather than *maximization*. In this manner, it would not be inappropriate to call green political economy the 'economics of enough'. It may be that many economies of scale turn out to be ecologically irrational.[13] One reason for this is that although a self-regulating market economy may deliver an optimum *allocation* of resources within the economy, it is unlikely to result in an optimum *scale* of the economy relative to its environment (Eckersley, 1991: 6). According to Daly and Cobb, 'Environmental degradation must be shown to result from the *scale of the economy in general*, rather than only from allocative mistakes that can be corrected while throughput continues to grow exponentially' (1990: 368, emphasis added). 'Economies of scale' may increase the scale of the economy beyond that which the environment can sustainably support. The problem with orthodox economic theory in this respect is that a Pareto-optimal allocation of resources does not say anything about the ecological sustainability of that scale of resource use.

Exporting pollution is less likely within a locally based economy, since the scale and type of technology and production processes used are governed by ecological and not just economic considerations. It is in the interest of the local (human) community to ensure that its economy does not undermine its ecological basis. According to Dryzek (1987: 225), local control over the economy–ecology metabolism is more efficient in terms of negative feedback than non-local control. Within the context of the present global economic system, the green local market economy perspective encourages local self-reliance as a positive benefit to be gained from delinking from the global economy (Morris, 1990; Sachs, 1990). Thus, this can be viewed as a 'bioregional' approach of 'greening the whole by greening the parts'.

From an economic point of view, by far the biggest distinction between local market economy and capitalist market economy is the greater immobility of capital. Paradoxically this assumption of capital immobility is at the heart of orthodox economic arguments for the specialization of production and international trade, as expressed in the law of comparative advantage. The problem is that capital is mobile at the global level, particularly since the deregulation of international finance in the early 1980s. The international mobility of capital means that 'investment is governed by absolute profitability and not by comparative advantage' (Daly and Cobb, 1990: 216). In encouraging capital immobility, so that industries as well as investment stay within the community, green political economy simply makes the assumptions of orthodox economic theory real rather than abstract. Given that capital in the real world is extremely mobile, this highlights the gap between orthodox theory and real-world practice. With capital immobility any trade that does take place, which from the ecological point of view should be progressively minimized as we move from local to regional to national and global levels, will be as a result of comparative advantage. This concurs with Adam Smith's argument that specialization of production is limited by the extent of the market; the more specialized production is, the larger the market and trade required to absorb it.

Specialization within the contemporary global market offers an extreme example of the dangers of over-dependence on trade. An economy which has placed a large part of its domestic capital into some specialized area of production is vulnerable to the vagaries of the global market to secure the goods and services it needs, as countless 'developing' countries found to their cost in the 1980s. This example works all the way down from national economies operating within the global market, to local economies operating within national markets. It is not specialization that is the problem but over-specialization that leads to over-dependence upon imports and/or a dependence upon attracting footloose international capital. The green argument is against excessive trade on ecological grounds (the environmental impact of transport), outlined above, but also on the grounds that trade decreases local economic self-determination, because an over-reliance on trade weakens the ability of the local economy to meet its own needs from within. Keynes's thinking on the location of production is close to the green position. According to him, 'Ideas, knowledge, science, hospitality, travel – these are the things which should of their nature be international. But let goods be homespun whenever it is reasonably and conveniently possible and, above all, let finance be primarily national' (in Morris, 1990: 195). Keynes's concern that the monetary dimension of the economy be localized is something with which green political economy fully agrees, carrying it further as in LETS by encouraging currency to be created and controlled at the local level of the community. The key to green political economy is its stress on decreasing, as far as possible, the gap between production and consumption, decreasing the need for extensive, globalized trade. Part of the green argument against global trade is that the existence of a global market with powerful economic actors such as multinationals, and institutions such as the World Bank, the International Monetary Fund and the General Agreement on Tariffs and Trade, mean that poor countries trade under disadvantageous conditions. Green political economy is thus not against trade, rather it is against unnecessary and forced trading relations.

Increased dependence upon trade and foreign investment is the price to be paid for enjoying the range of goods and services that access to the global market offers, given of course that one has 'effective demand', that is, purchasing power to buy these goods. The green alternative to the precarious benefits of this situation is increased economic self-reliance, where quality of life conceptions of welfare or well-being may compensate for a more limited range of goods and where increases in productivity may be 'cashed out' in more leisure time rather than increased wages.[14] And as argued below, from the green point of view, being locked into the global economy and dependence upon trade also implies weakening economic self-determination and cultural diversity.

It is important to point out that this view of the self-reliant economy differs from arguments for complete self-sufficiency or autarky. *Extensive* trade and exchange are discouraged, but, given the spread of resources on the planet, arguing for complete self-sufficiency would leave some resource-poor economies in a worse position than they need be in the absence of trade and redistribution. At the same time, green political economy as outlined here does not have an anti-materialist bias in the way that some green theories criticize the 'immorality' or 'spiritually corrupting' effects of material consumption. Martell's view that 'greens perhaps sometimes undervalue the extent to which material acquisition and consumption can be a source of personal fulfilment' and that 'advances in material standards of living are as likely as frugality to further intellectual and spiritual fulfilment' (1994: 49–50) is closer to the position being defended here than the anti-consumerism of some green positions. The case for decreasing dependence upon trade is not an attempt to smuggle in an anti-materialism at the level of green principle. It is not proposed, for example, to simply replace the 'consumer' with the 'citizen'. Rather its basis is to be found in the ecological arguments against trade and, as argued below, in the erosion of self-determination and a conception of liberty as independence, and finally in the balance between the claims of 'autonomy' and those of 'welfare'. The main point is that consumption as a mode of human being, acting and experience is not abolished as some necessary requirement of fulfilling green political aims.

Together with the argument for self-reliance, not being geared primarily towards capital accumulation means the imperative as well as the ability of the local economy to grow after the manner of current national economies is absent. If we also add to this the regulative role of the state, both local and central, in implementing environmental standards, there is a strong case to be made that economic decentralization of the type represented by local market economies may play a central part in ensuring both local and national sustainability. If the economy's ability to expand is limited by the extent of the market, in ecological terms the smaller the market the less likely it is that the economy will expand beyond its ecological parameters. Lessening dependence upon the whole world as one's 'ecological hinterland' implies a much closer link between economic activity and the ecological conditions which facilitate that activity. That is, the dependence of the economy on ecological goods and services is more visible owing to the shortened negative feedback relations, when the economy is embedded in local ecosystems rather than using the resources of distant ecosystems. However, the 'ecological basis' of the modern human economy lies somewhere between 'ecosystem' (local) and 'biosphere' (non-local), and therefore trade cannot be ruled out.

This self-limiting character of local market economies harks back to an earlier tradition of political economy associated with Aristotle. This refers to the distinction Aristotle made between chrematistics and oikonomia within political economy. *Chrematistics* is defined as that branch of political economy relating to the manipulation of property, wealth and currency so as to maximize short-term returns to the property owner. Chrematistics, in short, mistakes a means for an end, and according to Aristotle it is characteristic of this form of acquisition that 'there is no limit to the end it seeks; and the end it seeks is wealth of the sort we have mentioned [i.e. wealth in the form of currency] and the mere acquisition of money' (1948: 1257b). *Oikonomia*, by contrast, is, according to Aristotle, a more limited form of acquisition. Its central concern is the 'management of the household' geared towards long-term maintenance of the welfare for all household members. The limited nature of this form of acquisition is given by Aristotle thus: 'the amount of household property which suffices for a good life is not unlimited' (1948: 1236b). It is clear that what sustainable development requires is integrating the 'management of the household' with the 'economy of the household': that is, integrating economy and ecology. This division between chrematistics and oikonomia can be mapped on to the distinction being made here between the capitalist market system and local, non-capitalist economic organization in which market exchange plays a part. This distinction represents the separation of the economy and the 'economic motive' (economic rationality) from social relations and other forms of rationality (Polanyi, 1957: 54), most notably, in this case, ecological rationality (O'Neill, 1993: 169). In this distinction lies one of the principal origins of the 'disembedding' of the economy from the society it supports and within which it is located. However, the argument, following the logic of not wishing to simply replace 'consumer' with 'citizen' as the only 'ecologically virtuous' mode of being, is not to replace chrematistics with oikonomia, but rather to require that chrematistic economic activity not be completely self-regulating but, in part, be delimited by oikonomic norms.

The local market economy is one which is not just quantitatively different (in terms of overall ecological impact) but also qualitatively different from the contemporary capitalist market economy. As a market system where individual producers and consumers meet face-to-face, a local economy may lead to the sort of many-sided relations between community members of the type argued for by eco-anarchists, as discussed earlier. LETS offers one way in which the local market system may reintegrate the formal economy and the wider social system as a necessary prelude to the reharmonization of society and environment. Re-embedding the economy within society is thus a necessary step on the way to reintegrating the human economy and nature's economy, a central part of which requires integrating the formal and informal economies, and ultimately a reconceptualization of the 'economy'.

The local economy may also be said to display characteristics of a 'convivial economy' (Illich, 1975), a long-standing green view of a more 'human-scale' and sustainable economy, in which 'responsibly limited tools' and technology are used by people rather than vice versa. The economics of the local economy finds a counterpart in the 'soft technologies' of alternative energy production and the organization of economic production. It also resonates with the 'small is beautiful' philosophy associated with Schumacher (1973) and his ambition to create an 'economics as if people really mattered'. As Latouche has pointed out, 'One goal of many Green groups is to recreate a convivial society through deliberate construction of small-scale community and solidarity networks of all sorts' (1993: 237). Where this model of the local economy differs from Illich's 'convivial economy' is that although the economy as presented here includes 'activities of people when they are not motivated by thoughts of exchange . . . non-market related activities through which people satisfy everyday needs' (1981: 57), it also includes activities geared towards exchange, although as far as possible, within the local market as opposed to the global economy.[15] In terms of the distinction drawn in Figure 6.1 between the shadow, convivial and formal economies, the green local economy and green political economy represent a redefinition of the 'economic sphere'. On one level, the local economy can be located within the cash-based informal and non-cash informal economy. At another level, it may be argued to represent an alternative formal economy in that although it is geared towards exchange, it works with a different economic rationality

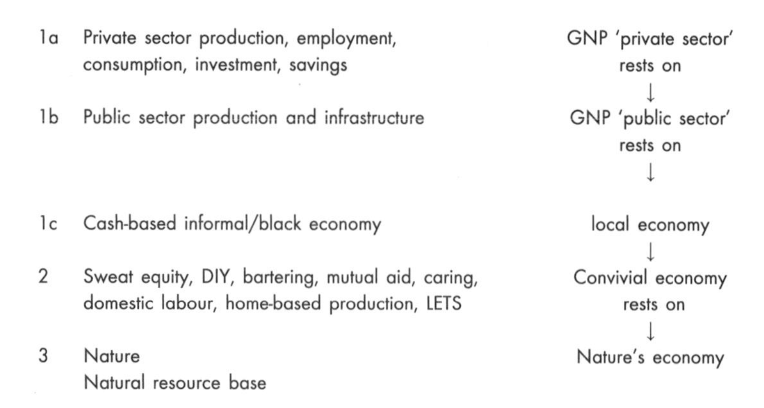

FIGURE 6.1 *Adaptation of Henderson et al.'s (1990: 267) model of the total productive system of an industrial society*

as well as a different currency and financial substructure than the formal or cash economy. It is above all, local.

From an ecological point of view, all purposive human activities which impact upon the environment can be considered as particular instances of the overall metabolism between the human and the natural economy. That is, from a green political economy perspective, the fact that one's activity does not command a price or is not registered within the formal market economy does not make it any the less an economic-ecological activity. The logic of this position leads to the reconceptualization of the 'human economy', its extension to include all purposive activity which impacts upon the environment as part of the metabolism between society and nature. Here I follow Hayward's suggestion that 'If a unified theory of economics and ecology is to be possible, it will neither hypostatize an opposition between economy and ecology nor posit a straightforward identity of the two' (1995: 116). Green political economy is an attempt to construct a unified theory of economics and ecology.[16] Its aim is to argue that the resolution of the ecological crisis demands a redefinition of what we mean by the 'human economy' in order that the latter be brought into harmony with the environment.

In just the same way that non-monetized activities are considered a legitimate part of the metabolism between economy and ecology, it also follows that such non-monetized activities cannot be viewed as 'unproductive' or valueless. As Hayward (1995), Mellor (1992; 1995; 1997), Salleh (1995; 1997) and other eco-feminist writers have argued, what is missing from orthodox theories of political economy is the whole realm of 'reproduction' as a site of valuable work and purposive activity. Contemporary orthodox political economies, whether liberal, Marxist or social democratic, identify production in such a way as to exclude types of labour and forms of intentional activity which are vital to the human economy as a whole, including that part coextensive with commodity exchange. Just as nature's economy would exist without the human economy (indeed it is the contention of some radical deep ecologists that nature would be better off without human activity), so green political economy holds that the human economy can exist without the money/formal/market economy. However, green political economy differs from those for which the only ecologically rational metabolism between society and environment is the complete rejection of the market economy and the return to a completely transparent 'convivial economy'. What green political economy attempts to highlight is the *desirability* (not the necessity) of a balance between the formal and informal parts of the human economy as a *necessary* part of creating a balance between the economy and the wider economy of nature. The reason for this is that the choice of a balance between the informal and formal economies is a matter of collective choice to be made under the rubric of 'freedom' rather than 'necessity'. That is, it is a democratic choice concerning the sort of society we wish to have and live in. If the 'choice' of the metabolism is under the rubric of necessity, there is a greater chance that this political choice may be 'technocratically constrained' and undemocratic.[17]

While orthodox theories of political economy focus almost exclusively on the formal or monetized economy, green political economy attempts to show the relationship between the 'formal' and 'informal' sectors of a society's economy. This aim of green political economy is concerned with demonstrating how the formal, monetized economy is dependent upon the non-monetized economy which in turn is dependent upon the environment. Henderson's model of the 'total productive system of an industrial society' (Figure 6.1) represents the main thrust of the green political economy perspective. In presenting and understanding the economy in this way, green political economy attempts to overcome what it perceives as the inadequacies of contemporary economic thought. These weaknesses include the model of the economy as a closed system, creating the illusion that the human economy is separate from the natural world and self-generating, just as markets are assumed to be self-regulating. Another standard green argument is the critique of orthodox economic measurements of social welfare/wealth, such as GNP (Eckersley, 1992b; 1993a), in favour of alternative qualitative indicators or the radical alternative of abandoning the search for indicators altogether. As such green political economy represents a demand for the radical reconceptualization of economic theory. To some extent, it is part of a search for a 'post-industrial' (but not necessarily anti-industrial, as argued below) political economy, premised on the critique of the industrial economics of contemporary economic theory. As such green political economy is consistent within Giarini's view that:

> 'economics' for the last two centuries has been the 'economics of industrialization' and not of the economy, which includes all assets and efforts that contribute to welfare . . . ecological and other current movements (e.g. women's liberation etc.) are all directed at the rehabilitation of non-monetized assets and activities which contribute to wealth and welfare and which have been marginalized or left out of account in the traditional economic (and socio-economic) system. (1980: 369)

The integration of economy and ecology requires a green political economy perspective which goes beyond the 'greening' of orthodox economic thought, as given by neoclassical environmental economics. The latter, while useful, does not capture the full *metabolic* character of the relationship between the human and the natural economies. Green political economy represents a move away from thinking about the metabolism with nature through monetized exchange of commodities on the formal market.

Money and Green Political Economy

Following the Aristotelian distinction between oikonomia and chrematistics, which gives rise to the 'household' and the 'market' as competing models of political economy, the LETS economy highlights a centrally important aspect of green political economy. This has to do with a critique of money in economic activity that has been developed by writers sympathetic to the green position, such as O'Neill (1993; 1995a; 1995b; 1995c), Lee (1989) and Altvater (1993). According to Aristotle, chrematistic activity is concerned with accumulating currency, the means of exchange, while 'economic acquisition, that of the household, considers acquisition only with respect to the object's primary use, as an object that satisfies a need' (O'Neill, 1995a: 426). There is a limit to the accumulation of such goods, according to Aristotle. On a 'householding' view of the economy, there are thus limits to wealth and property. This is not so with the acquisition of money, as Aristotle, Locke and early theorists of capitalism understood.

According to Lee (1989), the development of the money economy was central to the modern market economy, and laid the basis for the separation of the human economy from nature's economy. The separation of the economy from its ecological context also meant the increasing separation of the economy from wider non-economic considerations. In ecological terms, Locke's argument in defence of money (and the inequality that a money economy requires and justifies) permitted the accumulation process that is at the heart of the capitalist market system. Until the creation of money and its widespread acceptance, wealth and accumulation were limited by natural constraints (the limits of a person's stomach or the length of time natural products would last without spoiling). With the widespread use of money as a non-putrefying store of wealth, limits to accumulation could be overcome. As Gorz puts it, 'once you begin to measure wealth *in cash*, enough doesn't exist. Whatever the sum, it could always be larger' (1989: 112, emphasis in original). With money not just as a medium of exchange but now as a store of value, what Lee calls the 'organic' basis of human wealth was overcome. With the invention of money, 'Accumulation of this non-putrefying object on the part of the individual can now be limitless and go on for ever, the accumulation process having being emancipated from the workings of Nature' (Lee, 1989: 164).[18] In terms of Figure 6.1, it is money which is at the root of the separation of the formal economy from both the non-money economy and nature, the separation of economic from ecological rationality, and leads to a blindness with regard to the ecological origins of and contributions to human wealth.

While recognizing the benefits of money as a medium of exchange, the operation of LETS seeks to counteract the transformation of this medium into a store of value. LETS 'currency' is a medium of exchange. There is no benefit to be gained from accumulating it, since the sole function of a LETS unit of exchange is to facilitate trade and exchange. The difference between the contemporary cash economy and the LETS economy is that, in the latter, currency is purely a means of transmitting information to enable trade. That is why LETS is a non-monetary form or barter system of exchange but without the disadvantages of barter. 'Wealth' in a LETS economy is limited to the amount of goods and services one can trade for within the LETS scheme. Although the accumulation of goods bought within the LETS market is possible, this type of accumulation is quantitatively and qualitatively different from the type of accumulation that Locke sanctioned and Aristotle criticized. The LETS economy brings out the central aspect of green political economy's emphasis on 'use value' as constitutive of wealth in opposition to accumulation of 'exchange value' which is the understanding of wealth within the money economy and its various theories of political economy. To put it another way, trade and exchange within a LETS economy are geared towards the satisfaction of needs and wants facilitated by a medium of exchange, and are not concerned with accumulating the means of exchange and transforming it into a store of value. The use of goods and services to fulfil needs and wants is the principal object of those who engage in trade in the LETS economy; trade is not seen as a 'moment' in accumulating money. Within LETS, 'debit' is understood as a promise or a 'commitment' to render services or products to the same value some time in the future. Whereas in the formal economy one needs money to buy the things one needs, which means that one must either work or receive welfare payments, within a LETS economy individuals 'create money' in the act of exchange itself.[19]

Precedents for the green critique of the role of money within the contemporary economy, and in particular the disproportionate power of finance in affecting production decisions, can be found in the work of Douglas on 'social credit'. In part, Douglas's critique of production for production's sake, regardless of its social usefulness, is close to green concerns, as Hutchenson (1992) suggests. According to this line of argument, production is driven by the imperative to generate enough money in order that there be sufficient monetary means with which to purchase the goods and services required to fulfil needs. As Hutchenson puts it in explaining the rise of the money economy, 'Goods are produced as a means to an end – to secure the money with which to meet basic subsistence requirements since access to resources as a right from the commons has been denied through the enclosures' (1992: 6).[20] From Douglas's ideas on 'social credit' one can trace key aspects of green arguments for a guaranteed citizens' income. One of the main ecological reasons put forward for a citizens' income is that divorcing income from work would undermine the ecological irrationality of having to increase production in order to generate sufficient purchasing power for the distribution of existing supplies. The deregulation of currency along the lines of the LETS model, and the citizens' income, are two of the main financial proposals found within the economic policies proposed by greens.

This idea of increasing production in order to pay for the things we need underpins some of the reasoning behind ecological modernization, discussed in the last chapter. Ecological modernization is consistent with the view that financial resources are required in order to pay for environmental protection, as well as holding environmental standards (not to be confused with protection) as not a barrier to economic growth. Hence continued economic growth is a necessary prerequisite for achieving ecological sustainability. Rather than altering production and consumption patterns directly to be more ecologically rational, ecological modernization encourages us to think of ways to increase production and consumption so as to generate the necessary financial revenue with which to pay for environmental improvements. Economic growth, like Achilles's lance, is held to be able to heal the environmental wounds it inflicts. Now whether or not 'economic growth' (however this is defined) can be reconciled with ecological demands is a moot question, and there may be more in the ecological modernization idea of the ecological efficiency of *differentiated* growth than green critics allow (Jacobs, 1995; 1996). However, the main point here is that the elevation of economic growth, in terms of increased production and consumption, to the status of a 'given' is partly to be understood in terms of the central role of money within industrialized economic systems. In other words, it is not just the consumptionist lifestyle that drives the economic engine, but also a more fundamental relation between production and the generation of sufficient monetary purchasing power within the economy.

Virtue, Production and Consumption

While many green arguments concerning political economy are articulated in terms of economic–ecological relationships, there is also a moral dimension to green political economy. In this section I want to highlight those aspects of green political economy which pertain to virtue.

The goal of economic self-reliance is advocated as much as a moral ideal as it is a particular means by which a more ecologically rational economy–ecology metabolism may be established. Self-sufficiency or *autarkeia* was a virtue central to Stoic thought, for example, where it was understood in terms of detachment from worldly concerns and care only for the cultivation of individual virtue and rationality (Slote, 1993: 645). On this gloss, to be self-sufficient was to be untroubled by temporal matters and to devote one's attentions and energies to the important spiritual matters of life. Although this spiritualized account of self-sufficiency does find its defenders within green theory, particularly amongst those who decry the 'materialism' and 'consumerism' of modern societies, there is another related virtue, that of self-reliance, which is a more attractive ecological virtue. Whereas self-sufficiency as a virtue implies a notion of detachment, and an inward-looking, almost contemplative disposition of inner contentment, self-reliance does not imply any of these qualities, but rather denotes a sense of relying on one's own steam. Self-reliance conveys a sense of autonomy, independence and self-determination.

Self-reliance is understood both as a constitutive characteristic of communities and as a character trait of individuals. On the community level, self-reliance is argued to increase awareness of the dependency of the human economy upon the non-human world (Dryzek, 1987; Goldsmith et al., 1992). At the same time, self-reliance decreases dependence on imports and trade to provide those goods and services we think of as constitutive of the 'good life' for us (Allaby and Bunyard, 1980).[21] Thus the virtue of economic (and ecological) self-reliance is its enhancement of the capacity for self-determination. According to Benjamin Franklin, 'The man who would trade independence for security deserves to wind up with neither.' The green argument is that extensive trade increasingly erodes both independence and economic security. This is often the underlying argument of green opposition to extensive trade, and is one of the main arguments used in favour of delinking from the global economy. From a green point of view, participation in the global economy ought to be presented as a choice between economic welfare through trade, and independence through increased self-reliance. It is not simply the fact that extensive global trade has damaging environmental effects in terms of transporting goods, resources and people, and the infrastructure (roads in particular) required to facilitate it. At the same time, such patterns of trade within the global economy lead to domestic economies, and the societies they support, becoming over-dependent upon trade and having to fit into a global division of labour. These economic-structural imperatives weaken the capacity of domestic economies to determine their own development path. In terms of welfare and autonomy considerations, the green argument for self-reliance is an attempt to redress the balance in favour of autonomy while avoiding the possibly welfare reducing effects of complete self-sufficiency. What the argument for self-reliance comes down to is that collective self-determination within the global economy requires decreasing dependence on trade to fulfil collective welfare needs.

We can also conceive of this green argument for self-reliance as an argument for a particular conception of liberty. For example, according to Windass:

> a community which uses local materials and its own skills in order to house itself, clothe itself, and feed itself . . . is 'free' in a much more radical sense than a community which draws a large income from the industrial treadmill and spends it all to buy necessities from outside. (1976: 574)

While not completely agreeing with this, it is a good example of green arguments for self-reliance in terms of a particular conception of freedom as self-determination and autonomy. Here, as in other aspects, green political theory demonstrates a remarkable continuity with earlier classical and more modern republican concerns of the tension between liberty and luxury. Just as writers from Aristotle to Machiavelli, Rousseau and Arendt (Whiteside, 1994) drew attention to the negative impact of excessive economic wealth on political society, encouraging *inter alia* atomism, the elevation of consumption as the key component of the good life, a retreat into the private sphere, and dependence upon complete strangers and forces outside one's control, likewise green politics stresses the negative impact of trade within the context of a global market on both individual and collective self-determination. Greens argue for the moral as well as ecological benefits of self-reliance. As indicated earlier, part of the green argument is that beyond some threshold, the welfare benefits from active participation within the global market economy begin to undermine the conditions for autonomy, as well as the benefits themselves decreasing in their ability to deliver utility, or contribute to human welfare (Hirsch, 1977). Rather than seeing economy–ecology problems as stemming from a scarcity of resources, the moral underpinning of green political economy seeks to recast these problems as stemming, in part, from excess demand. In line with the classical view, green political economy holds that temperance and the avoidance of excess are virtues (Clark, 1994: 119), but adds that the satisfaction of those moderated wants should, as far as practicable, be in a manner which does not compromise self-determination.

Within the context of contemporary market economies, what greens are saying is neither complex nor all that new. People are offered a choice. On the one hand there is the current system of international trade, the global economy and economic interdependence. One radical green alternative suggests that people live more within their own local means and seek to provide the necessary goods and services they require without over-depending on external sources. This, in part, requires the moderation of desires, and redefining, but not rejecting, material consumption. This is the model of self-defined needs: we choose to moderate our desires for the sake of autonomy (Slote, 1993), while not construing the 'private sphere' as a sphere of privation. From a green point of view, autonomy is not equated merely with the satisfaction of preferences, but is also about ensuring autonomy in what Sunstein calls the 'processes of preference formation' (1995: 205), which he claims is a goal of democracy. This issue will be further discussed in the next chapter.

Also, as the classical economists noted, the scope of market exchange increases in line with people's desires beyond need fulfilment. The human capacity for food may be limited by the capacity of our stomachs, but the capacity for luxuries has no such 'natural limit'. Thus the more the demand for luxuries grows, the more the market (that is the economic wherewithal to supply that demand in terms of production and exchange) will grow. It is here that we come to the green argument concerning the inverse relationship between economic growth and a particular view of democracy, a view which in some ways echoes the misgivings of early commentators on capitalism concerning the detrimental effects of luxury on virtue. The more a society revolves around the pursuit of economic growth, the less is given to active citizen participation and involvement in the democratic life of the polity, as well as the democratic accountability of government, discussed in the next chapter. This observation is one that has been a constant critique of capitalist liberal democracy from Smith, Jefferson and Tocqueville to contemporary radical democrats and greens. It is also present in Aristotle's critique of *pleonexia*, incontinence or weakness of will.

In this way green anti-materialism does not stem from a moral rejection of materialistic lifestyles on the grounds that 'lives in the growth economy will tend away from the elegant and towards the grubby and materialistic' (Dobson, 1990: 88). The green critique of economic growth, materialism and consumerism also has to do with its conception of liberty as self-determination. Indeed, the critique of consumerism from the point of view of its negative effects on liberty as self-determination is stronger than its critique on the grounds that it falls short of a particular, and rather narrowly defined, 'green' conception of the good life. There is no reason to suggest that green politics necessarily requires a determinate or 'thick' 'view of the good', as opposed to a (limited) plurality or views of the good compatible with green aims and values.

Green political economy does not seek to reject individual material consumption as unworthy, or indeed as necessarily unecological. Stretton's (1976: 68) argument that one cannot enjoy common goods, such as public spaces, without some minimal degree of private, individual goods is an important one. Without private goods, according to him, individuals will use public ones inappropriately, i.e. degrade them. Although frugality and simplicity of lifestyle may have their own virtues, a *moderately* materialistic lifestyle need not be without its own inner rewards, and ecological virtue. But these virtues will not be realized unless consumption is integrated as an integral aspect of ecological stewardship. Material consumption becomes an ecological vice where it orientates itself around nothing other than consumption itself. The resolution of the tension between frugality and excessive consumption lies in an, admittedly imprecise, middle position of optimality or sufficiency. One can encourage 'voluntary simplicity' on ecological grounds, and on democratic grounds if the particular organization of the economy which provides that consumption has the potential to compromise democratic practice. However, it needs to be stressed that a reduction in consumption is not across the board but may be required in some areas (such as private car use). In other words, the green anti-consumptionist argument is a selective reduction of consumption within the context of reorganized production and consumption patterns rather than a *carte blanche* rejection of material consumption.[22] As Redclift (1996: 8) suggests, an attempt must be made to 'recover' consumption. Recovering consumption involves not just 'greening' it (as in purchasing 'eco-friendly' commodities), but also encouraging more durable, reusable commodities (challenging the 'throwaway' culture of consumption), as well as integrating consumption with other modes of human experience and action.

Allied to the green defence of economic self-reliance as a constitutive aspect of liberty as autonomy is an argument which sees work as intrinsically as well as instrumentally valuable.[23] The green conception of liberty is one which sees economic life and its concerns not as a precondition for autonomy, but rather as a potential site of autonomy in its own right. An example of this is Allaby and Bunyard's view of the moral benefits of having one's own power supply. For them:

> The family whose power supply comes from its own windmill or water turbine . . . feels it has *freed* itself from a situation in which the amount, form and method of supply of power is decided by an organization that presumes to know better than they what it is that they need and should have. *It is an escape from paternalism, a maturing.* (1980: 139, emphasis added)

It is not suggested that an over-riding and immediate green economic aim is to decentralize energy production to the household level. Rather, we should interpret Allaby and Bunyard's position in terms of the desirability of searching for alternative, less 'disempowering' ways in which to organize economic life, but searching for these alternatives does not automatically preclude the maintenance of existing forms of economic organization. This value may take a number of institutional forms, only one of which is that each household or community has its own (sustainable) energy supply. The main point is that economic self-reliance is a virtue in the sense that it is not only a means to an end (autonomy) but itself a part of that end. This goes hand in hand with the green rejection of the neoclassical classification of 'work' as 'negative utility' and engaged in primarily as a means to secure money.

In the global context of the contemporary world, what this notion of self-reliance attempts to express is that the increase in the distance between the point of production and the point of consumption is a good indication of the decrease in the capacity for individuals and collectives to be self-reliant economically. *A dominant principle of green political economy is thus to decrease the gap between production and consumption as much as possible.* This virtue of economic self-reliance is linked to the bioregional distinction between 'ecosystem' and 'biosphere' people discussed earlier. Ecosystem people, in being self-reliant (but not autarkic), are also ecologically rational. Being dependent upon a local ecosystem for the majority of the goods and services one consumes means that more attention will be paid to its health and continuing ability to supply those goods and services, including, most importantly, environmental goods and services that are essential to economic activity and human well-being. An ideal green economy might then be one where there was a market (with global, regional, national and local levels), but the majority of the things people required were produced either by themselves or at the local economy level. Of course not all goods and services can be produced locally, but the principle of appropriateness suggests that they should be produced at the lowest level possible. This would be in the spirit of the quote from Keynes above, which emphasized not only the local production of goods and services but also the trade and exchange of ideas and knowledge. It is not a contention of green political economy to constrain the production and free exchange of knowledge, information and science. As will be recalled, in Chapter 5 science was presented as a key form of knowledge necessary to cope with ecological problems, both locally and globally. However, one implication of the logic of self-reliance and an environmentally based distaste for global free trade is a more protectionist economic policy, as argued by Lang and Hines (1993).

At the individual level, the closing of the gap between production and consumption can be understood as an attempt to recapture some of the internal goods of work that are largely missing from modern forms of production.[24] One option is to alter or restructure current productive conditions so as to allow the internal goods of work to be potentially realized. This option includes worker participation schemes, flexi-time, multi-tasking, non-assembly-line forms of production and working from home. Such schemes can provide opportunities for the realization of some internal goods of work such as autonomy, solidarity, creativity and innovation, education, self-esteem and self-confidence.

Another way to realize the internal goods of work is by encouraging self-production both individually and collectively through economic practices such as LETS. Self-production, in weakening the link between money and production, opens up the possibility of production becoming a site of freedom and not simply a sphere of necessity. Part of the rationality of 'ownwork', to use Robertson's (1983) term, is to permit criteria other than 'maximization' and 'efficiency' into the sphere of production. As ownwork is not engaged in primarily as a means to secure money, it conforms to the ideal of a craft rather than an industrial mode of production. While not wishing to endorse all that Lee (1989) says on this issue, her basic point concerning the 'internal goods' of self-production, as opposed to the instrumental view of 'industrial production', is something that concurs with the spirit of green political economy. According to her, 'The pursuit of internal goods is said to constitute a morality of production or the artistic mode of production, while the pursuit of external material goods is said to constitute a morality of consumption' (1989: 222–3).[25] The understanding of 'work' is central to orthodox political economy and it is perhaps no coincidence that the common definition of work as 'disutility', something engaged in for monetary remuneration, is most marked in the Austrian school. Conceiving of work as disutility, to be entered into for monetary 'compensation' in the form of wages, further entrenches the central role of money within economic life. Not only is money, as Chesterton said, like a sixth sense necessary to make use of the other five, but added to this is the idea that money is the primary reason for engaging in productive activity. Indeed, as pointed out above, within orthodox economics only that which commands a monetary value, a price, is counted as 'productive'. Historically, the separation of people from resources, as a result of the enclosure of the commons, made the acquisition of money a prerequisite for fulfilling needs. At the same time, the idea of work as 'disutility' entered into for monetary reward represented another act of enclosure. This enclosure may be understood as the transformation of 'work' into 'labour': the aim of economic efficiency, coupled with the valuation of work primarily in monetary terms, implies that there is no point in reorganizing work which would permit the introduction of internal goods.[26] This second enclosure was the enclosure of the informal economy by the formal, as outlined earlier, in the sense that, as Illich (1978) contends, modernization means having to buy things and services you once did yourself. Within green political economy, self-production as an organizing principle of the economic life of society is an opportunity for the practice of virtue, the realization of internal goods.

At the same time, self-production within the context of economic self-reliance is also a matter of character formation. Self-production is consistent with a less 'consumerist' character typical of contemporary economic views of *Homo economicus*. Part of the green critique of a consumerist economy is that in being directed towards consumption, it undermines the internal goods of production. Roberts interprets this as implying that 'Self-fulfilment in working time . . . is one of the "goods" downgraded by consumerism' (1979: 44). On this view the 'productivism', so often criticized by greens, within contemporary forms of economic thinking is at root a critique of 'consumerism', as Lee (1989) argues. Thus one can acknowledge the virtues of consumption but also reject consumption as the dominant or over-riding aim of economic activity. In doing so, green political economy sets its face against orthodox economic theory, which sets a premium on consumption. As Hirsch points out, from the orthodox view, 'Economic growth . . . is interpreted as growth in the capacity of the economy to meet . . . individual and collective consumption demands' (1977: 18). Economic self-reliance is thus tied up with a shift to a less consumption-orientated economy, on the grounds that such an economy will allow a space for the virtues of production, including most importantly the opportunity for self-production. While consumption may have its own virtues as suggested above, the green case is that unless the consumer is also a producer, these virtues may become vices. The fate of the unemployed who consume but do not produce is a case in point. But, and this is the crux of the green argument, the increase in unemployment itself is a result of a consumption-orientated society. That is, where production is primarily regarded as instrumental to the process of consumption, consumption itself may be said to be one of the underlying causes of increasing formal unemployment. Ironically, as thinkers from Arendt to Gorz and Hirsch have pointed out, the logic of a consumption-driven economy is to produce a class of 'permanent consumers'. These are excluded from a society in which 'production' (understood as formal, paid work) is central to one's identity, participation and membership: hence the great socio-economic, cultural and psychological damage that unemployment causes. The green argument is for a redefinition of the economy away from a consumption-dominated one, and one in which there is a balance between the internal goods of production and the benefits of the existing system.

It is for this reason that LETS, as an example of a form of productive economic activity, offers what many consider to be an ideal solution to unemployment in contemporary industrial societies. Such forms of activity do provide consumers with the opportunity to become producers, engage in productive economic activity, but without having to secure formal 'employment'. Some have argued that one of the most significant advantages of LETS-type activity is that it offers the unemployed the opportunity to become a contributing member of society, a 'full citizen' as it were (Barry and Proops, 1996). Participating in such informal activity helps overcome the social exclusion and dependency upon welfare benefits which are among the most damaging effects of formal unemployment. However, given that the values that LETS activity creates are not commensurate with those that dominate the larger, formal economy, the contribution it can make to overcoming social exclusion is dependent upon some other form of public recognition of its 'economic and social standing'. In part this is the reason for expanding the notion of the 'economy' to include such informal activities, so that at least conceptually these productive and socially useful activities are recognized as integral aspects of the human economy. If, as many predict, the 'post-industrial' future of contemporary societies will consist in an accelerated decrease in formal employment, then LETS-type informal economy activity can become a key aspect of a future economy, if we wish to avoid creating a permanent group of consumers and socially excluded citizens. LETS can thus be viewed as a complementary (or alternative) way of resolving the standard conflict between 'environmental protection' and 'employment' to that suggested by ecological modernization. If the sphere of socially valued productive work is expanded to the informal or social economy, then unemployment in the formal sector need not imply becoming dependent on the state to support 'compulsory consumption' via unemployment and welfare benefits.

In this reorientation of the economy, a different set of character traits may be fostered, and ones that can affect production more generally and not just self-production. That is, production within the context of self-reliance, where it is not engaged in primarily for reasons of extensive trade for profit maximization within the global market, can be governed by internal rather than external criteria. The organization of production within a global economy is different from the organization of production within a self-reliant economy. In short, it is more likely that production as, at least in part, a social practice will be fostered within a self-reliant economic context rather than within a global economic one.

Another, more standard critique of consumerism, which can be found in a diverse range of thinkers from Aristotle to Smith, Arendt, Marcuse and Fromm, is that it promotes a view of individual well-being in which 'consuming' rather than 'doing' or 'being' is central. The point is not that consumption has no part to play in cultivating ecological habits, or a green view of the good, but that consumption within an economy simply geared towards consumption, with no recognition of the internal goods of production, makes for the cultivation of unecological characters.[27] Again it must be stressed that this is not the standard anti-materialist argument that is found in the many green critiques of contemporary society. Rather this position holds that consumption can become an ecological vice if it is not integrated within a mode of behaviour which recognizes limits to consumption. The point is that limiting consumption is not done for its own sake, but rather it forms a part of a mode of action and character in which the goods of consumption are not to become vices by being pursued at the expense of other goods. In terms of character, a strong claim is that consumption of the world, divorced from other human interests, consumes the world even as it fails to recognize the dependence of consumption on the world. Consumption as a mode of interaction with the world, under modern conditions where there is an increasing gap between production and consumption (as a result of such processes as the social and global division of labour and unemployment), is a permanent danger unless it is integrated within a wider mode which can prevent it from becoming an ecological vice. This wider mode is that of stewardship, expressed through the practices and roles of ecologically responsible citizenship, and sustainable and/or symbiotic production or work (including self-production). While not wishing to return to the type of metabolism with nature that was criticized in Chapter 2 (in which we simply 'followed' nature), we have to acknowledge the *metabolic* character of social–environmental interaction. A consumption-driven and -centred economy, particularly within the context of a global economy, in which consumption is placed within a biospheric as opposed to an ecospheric context, can create the impression that the human economy is 'closed' or independent from ecological limits, as suggested earlier in the critique of orthodox neoclassical economic models. Limits on consumption can of course be imposed, but the prescriptive aim of green political economy is to suggest *responsible consumption* (not just 'green consumerism') as a valued activity and basis for personal identity and privacy. Consumption must be prevented from being taken to excess and becoming an ecological vice, not simply dismissed as intrinsically unworthy and necessarily unecological.

Temperance in the acquisition of material goods is a more accurate and attractive interpretation of the green argument against consumerism than its condemnation on the grounds of its intrinsic worthlessness. By couching the green argument for self-reliance and self-production in terms of moderation one avoids the popular parody of the green position as essentially a matter of ascetic self-denial and sacrifice (Allison, 1991). It can also be used to supplement the common green view that ecological problems are as a result of supply-side scarcities (resources and sinks). The critique of consumerism views the problem as also due to excess demand, which therefore cannot be overcome by 'supply-side' economic solutions. Thus the green critique of *undifferentiated* economic growth can be linked with Hirsch's (1977) analysis of the 'social limits to growth' (Barry, 1990).

Green political economy does criticize the excesses that characterize the consumer ideal, as well as the instrumentalization of production (as a result of specialization and the demands of the division of labour) that this ideal requires.[28] However, the alternative is not the abolition of consumption and the benefits derived from it, but rather the integration of consumption with production on the basis of self-reliance and moderation. The essential positive values underlying consumption, that is, the desire to live a little better, its 'civilizing' effects, and its contribution to personal identity, need to be 'liberated' from its current conceptualization which divorces consumption from production. It is the problems that arise from too great a separation of consumption from production, in both physical and conceptual terms, that explain why green political economy seeks to close the gap by integrating them as much as possible.

Decreasing the discontinuity between production and consumption, increases the chances of securing the virtues of consumption. That is, the civilizing, enriching and broadening effects of consumption, noted from Smith to Marx to modern liberals, demand a reciprocal relationship with production, which in turn requires a (largely asymmetric) relationship with the environment. Where green politics differs from other political theories such as liberalism or socialism is that whereas the latter view the link between production and consumption in terms of ensuring full employment in the formal economy, the green view is to encourage an ideal of self-provisioning, both individually and collectively, within the informal economy, as much as possible, and restructuring the 'formal' productive sphere so as to enhance the internal goods of work. Clearly not all needs and wants can be satisfied in this manner, but the green argument is that this sphere of productive activity should be recognized as forming a significant part of the human economy and its metabolism with nature, and that this informal economic sphere should be protected from the colonizing tendencies of the formal sphere which seek to destroy it. Whether or not to work in the formal economic sphere should, on a radical interpretation of green political economy, be a voluntary not a compulsory decision.[29] In other words, the choice should be *whether* to work in the formal economy rather than *where* to work in it.

The ideal-type model for green political economy is that of the 'household', and informal sectors of the economy working in conjunction with the formal market or cash economy. As O'Neill (1993: 172) argues, Aristotle's conception of the household as the model for political economy is one that concurs with many aspects of green political economy. In terms of the argument about decreasing the distance between production and consumption, the ideal of householding is to reverse the current situation where, according to Illich, the household becomes the sphere of 'compulsory consumption' (1981: 112). Here the distinction between self-sufficiency and self-reliance is particularly important, for the aim of green political economy is not to make the domestic sphere a site of 'compulsory production'. As such there will be a continuing role for the formal economy, but one in which it is not automatically the only, or the dominant/visible, site for meeting needs.

For green political economy, the 'means of production' go beyond the traditional definition of industrial plant, machinery, specialized knowledge and the division of labour to include the domestic sphere as a site of production as well as consumption and reproduction.[30] Within green political economy the aim is to facilitate as many modes of production (consistent with ecological modes of interaction) as required, rather than imposing one: the industrial one. At the same time, and once again going against the grain of radical green thought, green political economy does not seek to abolish the industrial mode of production. Rather, as Illich argues, what is aimed for is 'technological maturity', a situation where, 'the industrial mode of production *complements* other autonomous forms of production' (1974: 86, emphasis added).[31]

A related point has to do with the conception of 'development' within green political economy. In contrast to orthodox political economies, green political economy does not prescribe a particular view of 'development' or 'progress', but rather sees this as one of the most important expressions of communal autonomy, the right of communities to decide their own understanding of social and economic development. From the perspective of many radical and Third World ecologists, a primary reason for delinking from the world economy is to prevent the imposition of the Western model of socio-economic development on communities throughout the world (Illich, 1974; Shiva, 1988; Sachs, 1990; 1995; Latouche, 1993). According to Latouche the creation of the global market system can be seen as the 'westernization of the world' (1993: 160), the hegemony not just of a particular economic system but also of a particular worldview and ethos. For many radical greens cultural autonomy is threatened by the global market economy. In the words of Wolfgang Sachs, 'No country today seems to be capable of controlling its own development' (1990: 336). The same global economic system that is threatening global biodiversity is also held to threaten global cultural diversity. The identification of the Western model of development with progress signifies an inevitability and desirability which is used to silence any criticism. Progress is good, and after all, you can't stop progress. A full understanding of the relationship between the virtues of self-reliance and autonomy would require investigating the notion of self-defined needs at the individual level, and 'chosen' rather than 'imposed' development paths at the collective one.[32] Thus the ecological virtue of self-reliance has to do with the relationship between 'autonomy' and 'welfare', in which welfare considerations relate to moderated and self-defined needs being fulfilled as much as possible by agents themselves.

Self-reliance is to be understood as a virtue because it stands as a mean between full self-sufficiency or autarky, and complete dependence.

Self-reliance as an economic aim attempts to acknowledge the positive role market and other forms of exchange may play in expanding the horizons of individuals and communities, while seeking to ensure that trade is the instrument of the community rather than the other way round.

Conclusion

The creation of a market economy represents the separation of economic rationality from other forms of rationality. Polanyi (1957), in his study of the emergence of the modern market economy as a distinct and self-governing subsystem, reminds us of the uniqueness of the modern economy by contrasting it with the role of the economy previous to the emergence of the market economy.

> For, if one conclusion stands out more clearly than another from the recent study of early societies it is the changelessness of man as a social being. His natural endowments reappear with a remarkable constancy in societies of all times and places; and the necessary preconditions of the survival of human society to be immutably the same. The outstanding discovery of recent historical and anthropological research is that *man's economy, as a rule, is submerged in his social relationships*. (1957: 46, emphasis added)

This aim of 're-embedding' the economy within society as a necessary part of reintegrating economy and environment is the aim of green political economy. However, collective ecological management of the economy–ecology metabolism requires attention to be paid to specifically economic issues, such as the role and place of money within the formal economy. The material metabolism between economy and ecology is as much dependent upon the central role of finance, and its influence over the production decisions within the formal economy, as upon the technology of production.[33] This aspect of green political economy has been expressed by Altvater as implying that 'Today the further evolution of society is possible only if the economic rationality of market procedures is firmly embedded in a complex system of social, non-market regulation of money and nature' (1993: 260). The re-embedding of the formal economy within nature requires re-embedding it, as far as possible, within parameters set by social, non-market norms, and politically governing it in a manner suggested in the last chapter. A clear, if extreme, example of this is Latouche's celebration of the 'informal economy' as the 'reinsertion of the economic within the larger social texture of life, to the point that sometimes the economic is *completely absorbed* within this texture' (1993: 127, emphasis added). However, the complete absorption of the economic by the social is neither necessary nor desirable. The point after all is not to dissolve economic rationality and the formal economy completely, but to find their appropriate places within a more ecologically rational mode of production.

Green political economy differs from contemporary forms of political economy in a number of significant ways. Firstly, it includes a central space for institutions and practices in the regulation of the economy and the behaviour of economic agents, and adopts an explicitly institutional approach to economic analysis. Examples include non-money, market organization of the economy, such as LETS, and commons regimes. A final concern of this institutional approach is to focus on the immoderation of desires (as ecological vices) and, as a possible cause of environmental problems, to focus on excess demand as much as on supply-side shortages.

Secondly, green political economy questions the ends of development rather than simply querying the means. It has no one particular model of development, in the same way orthodox political economy has a particular conception of 'economic modernization' as an animating principle. This is clearly related to the critique of consumerism, and the argument for production and work to be seen as more than an instrumental stage in the process of consumption. At its most radical green political economy is a political economy of 'post-development' (Latouche, 1993), and is *post-* rather than *anti*-industrial. Thirdly, the normative dimension of green political economy extends to the thorny question of what is to count as a 'resource' and what is to be a 'proscribed resource'. What this expresses is that within green political economy, 'resources' are not exogenously 'given'. Rather the metabolism between economy and ecology is to be governed by considerations of human interests alongside those of ecological sustainability and ethical symbiosis.

Fourthly, green political economy has a radically different conception of the 'economy'. Whereas orthodox political economy analyses the formal or money economy, green political economy has a much wider understanding of the economy, one in which the informal, non-monetized economy is important. Thus while green politics seeks to responsibly limit economic rationality, it also sees this as occurring simultaneously with a reconceptualization of economic theory. This reconceptualization demands a transformation not simply in the content of economic theory, but also in the discipline of economics, away from abstract model-building and its self-perception as a 'hard science' and back to its roots in political economy.

A key part of this desire to expand the 'economy', and which relates to the normative concerns of green political economy, has to do with shifting economic analysis and organization away from a consumption-driven, and consumption-centred, economy. While not rejecting the consumption of goods and services as a positive mode of human activity, my concern in this chapter has been to suggest that without a counterbalancing with other modes such as production, its lack of self-limitation leads to its tendency to become an ecological vice. Consumption needs to be integrated within a more expansive mode of acting and experience which recognizes the dependence of human productive activity on the creation and maintenance of a stable metabolism with nature. A consumption-driven economy, key features of which are production for the sake of consumption and the centrality of money, disembeds the economy from its ecological context.

While some greens seek to return to past economic arrangements such as commons regimes, green political economy represents a more feasible approach to regulating economy–ecology exchanges. This approach includes the reorganization of formal work to permit internal goods of production to be realized, self-production, LETS, the possible changes that a shift from a biosphere to an ecosphere economy would have on production, together with more recent developments such as the Local Agenda 21 process.

A necessary aspect of this return to political economy, particularly given the normative facets of green political economy, is the role of democratic decision-making and democratic accountability in 'governing' the economy–ecology metabolism. It is to the relationship between green politics as collective ecological management and democracy and the delineation of a green theory of democracy that we turn next.

Notes

1 The Pareto criterion has conservative implications since it rules out any allocative pattern in which some would lose and others gain. This principle can thus be seen as an important part of the growth imperative at the heart of orthodox economic theory. Economic growth, as opposed to redistribution from the rich to the poor, satisfies the Pareto criterion that there is no alternative allocative pattern in which everybody gains. This criterion thus constitutes one of the primary stumbling blocks against redistributive arguments. The relationship between justifications of economic inequality and economic growth will be discussed in the next chapter.

2 I owe this point to John Proops.

3 Pearce (1992) defends the neoclassical economic approach to environmental decision-making precisely on the grounds that it presents environmental issues in a form (i.e. in economic terms) which existing institutions of public policy-making can readily and easily assimilate. As he puts it, 'Defending the environment means presenting the arguments in terms of units that politicians understand' (1992: 8). This brings out clearly the political institutional context of neoclassical economics, and how one cannot understand the latter without being sensitive to the institutional background against which it takes place.
4 For a discussion of the Wise Use movement see Dryzek (1997) and Harvey (1996).
5 The issue seems to be between mediated responsibility for or *control over* ecological goods and services, and risks, and transparent, immediate relationships to them via *ownership*. Given the urban character of modern societies, it is clear that a collective ecological management based on mediated control rather than direct ownership has definite advantages.
6 However, there are those, such as the The Land Is Ours campaign in Britain, who challenge popular exclusion from the land and the decision-making processes pertaining to it.
7 The political theory underlying this 'homesteading' model of environmental management, which focuses on the metabolic relationship between ecology and agriculture (rather than ecology and economy), is a Jeffersonian-cum-Rousseauian vision of agrarian democracy. According to Jefferson, 'cultivators of the earth are the most virtuous and independent of citizens', adding, in a statement which echoes the anti-urbanism and anti-commercialism of some green thinking, that, 'Merchants have no country. The mere spot they stand on does not constitute so strong an attachment as that from which they draw their gains' (in Miller, 1988: 207, 210–11). As for some greens, Jefferson thought that to dwell in the country was to dwell in virtue, while living in the city was to risk corruption (Rennie-Short, 1991). For green defences of urban living, see Bookchin (1992a), Paehlke (1989) and Ferris (1992).
8 In the end one must conclude that free market environmentalism is transparently ideological, the latest phase of the right-wing libertarian political project. It is a reaction to the 'statist' implications of neoclassical environmental economics and any other solution to environmental problems in which the state has a central role or in which the market does not. This view is extended to the 'green movement' as a whole, which is perceived as a bulwark against the privatization of environmental resources, another form of social resistance to the extension of the discipline and advantages of the free market. For most free market environmentalists, greens, whether deep or shallow, are simply another pressure group using the political process to undermine the advance of market principles. Seen within the historical context of the collapse of communism and the crisis within the Western left, environmentalism is portrayed as an alternative legitimation for socialist collectivism (Cooper, 1989; Anderson and Leal, 1991; North, 1995; Ridley, 1995). Indeed, if the arguments concerning the centrality of democratic planning and regulation to collective ecological management of the previous chapter are right, then it may be that ecological issues can serve to relegitimize aspects of socialist politics.

9 The reason why green political economy is self-consciously normative lies partly in the fact that it is premised on the idea that humans have no 'natural' niche (Chapter 5). If we add to the fact that humans create their own niche, then it becomes possible to see the moral and political dilemma that shadows green politics in all its starkness. This dilemma consists in seeing that almost everything on the planet (and other planets) can potentially become a resource, given human technological capacities. When science and technology proceed to transform previous non-resources into potential resources, the moral question is whether to place the latter into the category of 'proscribed resources', or forgo the development and maintain the 'non-resource' status of the natural entity. This raises different issues from debates concerning turning current environmental resources into 'proscribed resources', such as vegetarianism and wilderness protection. Practically speaking, it may be easier to proscribe potential rather than current resources, since the former have yet to become embedded within particular ways of life. However, it would be a mistake to think that the only arguments in favour of permitting non-resources to be used as resources, such as genetic material, are predominantly economic, although this characterizes most cases. There are also medical, scientific and other serious human interests at stake.

10 In many respects, modern executives can be seen as the real decision-making body in contemporary liberal democracies, while legislatures (and the civil service) approximate decision-recommending bodies. I owe this point to Rosemary O'Kane.

11 It is worth pointing out that this has much in common with the bioregional distinction between 'ecosphere' and 'biosphere' perspectives and a central claim of social ecology, concerning the importance of economic self-reliance for political autonomy, mentioned in Chapter 4. The non-capitalist market as a central part of local economic autonomy and self-reliance can be seen as a decoupling of the local economy from the global economy: that is, moving from a biospheric model of the relationship between the economy and the environment to an ecospheric model.

12 Although LETS and other aspects of what one might call the green local market economy have evolved and presently exist alongside the formal market system, there is a debate as to whether they supplement or actively undermine the formal market economy. The question of whether a LETS economy could exist without the formal market is not something that is addressed here. See Barry and Proops (1996) for further discussion of LETS. LETS may be viewed as both an alternative and a complement to the current global market: as the local economy becomes more established in meeting people's needs, the less dependent individuals will be upon the global economy. Therefore while it is not envisaged that the present global market economy will be abolished, the logic of the green local market economy is for the global economy to shrink in significance over time, as a necessary aspect of environmental sustainability.

13 It may be that under conditions of uncertainty and ignorance, as Faber et al. (1992) indicate is the epistemological position we stand in relation to nature, optimizing returns rather than seeking to mazimize them is more rational. The latter is in keeping with Rawls's (1972) 'maximin' principle and would also seem consistent with the precautionary principle. Alternatively, it may

well be that not taking advantages of economies of scale for some productive processes may be ecologically irrational.

14 One of the standard observations of worker-owned firms and co-operatives has been that after a threshold income level has been reached, increases in productivity lead to a reduction in working hours. In neoclassical economics this is known as the 'backward bending labour supply curve', where after some threshold income level has been reached, increases in productivity lead to more leisure (i.e. less working time) rather than increased production or income. As argued in Chapter 7, this economic view is premised on an instrumental view of work engaged in for monetary reward. For an examination of the relationship between worker co-operatives and green theory, see Carter (1996).

15 Illich coins the term 'shadow economy' to describe those activities 'which support the formal economy not social subsistence' (1981: 100), while the convivial economy comprises activities which are geared towards social subsistence. A fully convivial economy is perhaps only possible if the economy has not undergone modernization; hence Illich's concern with warning developing countries about the dangers of what he calls the 'modernization of poverty' and the 'radical monopolization of needs' that orthodox, Western, industrial modernization brings.

16 This search for a unified theory which would integrate economics and ecology has long been the goal of many early ecological economists, from Boulding (1966) and Daly (1973a) to Georgescu-Roegen (1971; 1976). The roots of this unified theory may be found as much in the application of economic ideas and concepts such as production, consumption, exchange and labour to the workings of the natural world, as to the more general ecological idea that the human economy be viewed as dependently embedded within the wider economy of nature. In its early development, as Worster (1994: 291–4) points out, the science of ecology was viewed as the application of economic analysis to nature. This one-sided influence of economics on ecology had to wait until the 1950s and 1960s for the reverse, i.e. the application of ecological analysis to economics, to occur.

17 While accepting the argument that the formal economy must be maintained politically, so as to prevent it undermining the natural and social conditions of the human economy, it is not the case that concerns of the human economy become coextensive with politics. There needs to be a separation as well as a connection between the economy and the polity.

18 It is of course no coincidence that the Lockeian defence of money as a store of wealth arose at the historical transition from an agrarian economy to a commercial economy. The inherent self-limiting features of an agrarian society, in terms of its 'organic' conception of wealth and acquisition, as well as its obvious 'back to nature' and 'householding' qualities, may account for the predilection for a return to this type of society within some strands of green political theory.

19 This seemingly odd statement brings out clearly the idea of LETS as a 'credit-barter' system rather than as simply an alternative money system. One can write a cheque in the local currency to pay for one's purchases without having to have 'sufficient funds' in one's LETS account. That is, one can go into 'commitment' on the understanding that one will render an equivalent value of goods or services in the future. And by going into

commitment, one has effectively 'created' LETS money, in the form of the cheque lodged in the account of the seller. Essentially one has 'bartered' *future* goods or services that one intends to sell, in exchange for *present* goods or services purchased.

20 The money economy and the enclosure of the commons are thus two sides of the same coin: they are, from a green point of view, potentially negative sides of 'modernization'. I say 'potentially' in order to distinguish the argument being defended here from arguments which portray the enclosure of common resources as always and everywhere ecologically destructive and morally unjustifiable.

21 This awareness of our dependency is from the green perspective itself a virtue, which helps to mitigate the tendency of humans to think they are independent of the natural world or natural limits. An added point is that conscious recognition of human nature, of its vulnerability and of our needy constitution opens the way to feminist contributions to ecological thought, such as the extension of the 'ethic of care' to ecological politics. These relate to such central political concepts as autonomy, the economy, progress and production, some of which are considered below and in the preceding chapters.

22 One reason why it is important to stress the acceptance of individual consumption as a 'good' on principle is to dispel any (liberal) suspicion of the erosion of the private sphere. The democratization of the family is not sought, at least not at the intimate, individual level. However, given the green concern with per capita consumption, macro-level regulation of reproduction is a legitimate policy area for democratic decision-making. Taxes and subsidies, as well as the promotion and spread of economic security (as argued in Chapter 7), can help the 'demographic transition', and achieve a lower rate of population growth.

23 Lasch's 'populist producerism' which aims for the 'rehabilitation of work, not the democratization of consumption' (in Holmes, 1993: 134) is close to the position being developed here. However, this concern with revaluing work as intrinsically valuable is to be distinguished from the overlap between Lasch's critique of contemporary American society and the 'deep green' one. Both share a nostalgic yearning for a pre-industrial rural society, where the 'vices' of the city, of industrial progress and of affluence are held in check by the virtues of a society of independent household farmers and independent craft workers. Indeed, many deep greens take Lasch's complaint about the 'enfeebling of character' as a result of consumerism as an argument for recapturing the values of (if not returning to) a hunter-gatherer culture (Shepard, 1993).

24 It is interesting to note that concerns with the virtues of self-production arise at a time when work is disappearing from the work-based society, in the same way that the origins of an aesthetic appreciation of nature arose historically precisely at the time when human ability to destroy it also arose. As Williams has pointed out, 'An artistic reaffirmation of the separateness and fearfulness of nature became appropriate at the point at which for the first time the prospect of an ever-increasing control of it became obvious' (1992: 67). However, the difference in the case of the reaffirmation of the virtues and intrinsic values of self-production is that what is disappearing is formal paid employment, not 'work' as a purposive, transformative activity.

In this sense, the self-production ideal is a 'post full employment' phenomenon. On post full employment as part of green sociological theory see Gorz (1983; 1989), Keane and Owens (1986), Robertson (1983; 1985) and Barry (1998a).

25 The craft ideal of work and productive activity is not exclusive to greens but can be found in utopian socialism, particular the strongly 'aesthetic' versions of Morris, Ruskin and Wilde for example.

26 Since work as labour was conceptually viewed as 'disutility', a necessary evil, one would have thought that the economic aim was to minimize work by improving efficiency of production. Historically, as many studies have shown, productive efficiency as a result of technological improvements or changes in the organization of labour has led to a decrease in the hours worked much lower than what could be achieved given the level of technological and other productive improvements (Hirsch, 1977; Gorz, 1983). On the whole, productive improvements have resulted in fewer people working, rather than less work per person (Gorz, 1983).

27 As discussed in the next chapter, consumption is not simply a private activity, but also a social value. Thus, the green position is not simply a matter of rejecting consumption for either production or citizenship, but rather one of finding the balance between consumption and production, a balance which can only be created by, and maintained, but not exclusively, through the activity of democratic citizenship itself.

28 A common observation of the recent history of Western economies is that as the economy grows, the 'natural rate of unemployment' (basically the maximum level of unemployment consistent with full employment of all factors of production) also rises. That this has risen over the last three decades simply means that a successful economy requires fewer and fewer people working. That is, we increasingly have the phenomenon of 'jobless' economic growth. But at the same time, as Hirsch (1977: 41–51) points out, of the work that is available, there is a noticeable dearth of fulfilling, intrinsically rewarding and well-paid occupations.

29 One of the policy suggestions put forward by greens (but also endorsed by others: Van Parijis, 1992; White, 1995), is the guaranteed basic income scheme mentioned earlier (Dobson, 1990: 112–13; Kemball-Cook et al., 1991: 19–23; Eckersley, 1992a: 143; Goodin, 1992: 197–8). The aim of this policy is to decrease people's dependence upon paid work in the formal economy and to increase their opportunities to engage in informal economic activity and 'ownwork' (Robertson, 1985: x). This is discussed in more detail in the next chapter.

30 Toffler's (1970) idea of a 'prosumer' sphere of economic activity, where producers consume what they produce and produce what they need, fits with green economic and moral thinking.

31 Illich's vision of a convivial economy, in which engineered artefacts are geared towards 'more effective use-value generation', and where there is an equal 'right to access to raw materials, tools and utilities' (1977: 94, 95), is a radical version of the aims of green political economy: the creation of a socially embedded, ecologically sustainable and ethically symbiotic economy.

32 Gorz uses the language of rationality to express the same point when he claims that, 'Economic rationality is not applied when people are free to

decide their own level of need and their own level of effort' (1989: 111). When people are not free to decide their own development paths, as many greens and Third World activists maintain is the case within the context of the global economy, (Western) economic rationality determines 'progress' and 'development' (Shiva, 1988; Latouche, 1993: 136).

33 This interconnection between financial and natural systems leads Harvey to posit that, 'Money and commodity movements, for example, have to be regarded as fundamental to contemporary ecosystems' (1993: 28). In other words it is not simply energy and material flows that affect ecosystem development, because ecosystems exist in relation to specific ecology–economy metabolisms.

7

Green Politics and Democracy: Green Citizenship and Ecological Stewardship

CONTENTS

This chapter looks at the relationship between democracy and green political theory in general and in particular green democratic theory and practice. In much the same way as Kymlicka (1990) argues that any plausible modern political theory embodies a commitment to the view of individuals as deserving of equal respect and concern, one can posit democracy as a value to be considered as an essential part of all acceptable political theories. In this respect, green politics is no different in its claim to be part of the 'democratic project'. However, beyond a shared commitment to democracy, political theories differ as to *what* they understand by democracy, the reasons *why* they advocate it, and *how* they envisage its institutionalization. Although all theories worthy of respect and serious consideration endorse the general *concept* of democracy, they disagree over the different possible *conceptualizations* of democracy. On both these points, the concept and the conceptualization of democracy, questions have been raised as to their necessary connection to green politics.

It needs to be pointed out that in their practical political activity, environmental groups and the green movement have been at the forefront of efforts to 'democratize' state institutions, and the creation of more democratic and accountable forms of environmental decision-making. Examples include green efforts to open up access to information, particularly scientific data, and creating more open forms of public policy-making. This anti-bureaucratic strand of green thought also calls into question aspects of the ecological modernization model, outlined in Chapter 5, with its bureaucratic-corporatist overtones. According to Paehlke, part of this green suspicion of bureaucracy stems from greens' concern for 'greater openness and greater public involvement in administrative decision making' (1988: 295). In this respect we can tentatively conclude that the *practice* of environmental politics over the last 30 years or so seems to provide evidence for Paehlke's argument that 'an answer to future economic, environmental and resource problems [may be found] in *more* rather than *less* democracy' (1988: 294, emphasis in original). The green movement's emphasis on grassroots activism, 'bottom-up' organizational principles and what Doherty has called its 'complicated democratic project' (1992: 102) are further evidence of the democratic credentials of green political practice. In a sense then, although there may be a question as to the strict *theoretical* relationship between green political theory and democracy, in practice this tension is often more apparent than real.

The Eco-authoritarian Argument

A good starting point to explore the relationship between democracy and green politics is Tocqueville's suggestion that:

> General prosperity is favourable to the stability of all governments, but more particularly of a democratic one, which depends upon the will of the majority, and especially upon the will of that portion of the community which is most exposed to want. When the people rule, they must be rendered happy or they will overturn the state: and misery stimulates them to those excesses to which ambition rouses kings. (1956: 129–30)

This assumption of the positive correlation between material affluence and the stability of a democratic political order is one which is closely associated with 'modern' political traditions such as liberalism and Marxism.[1] In this section it is the negative corollary of this assumption, i.e. that material scarcity creates the conditions for political instability and a shift to authoritarianism, that will be examined. What can be called a 'Hobbes–Malthus' position underpins the 'eco-authoritarian' school of green thought, which in the literature is most closely associated with Ophuls (1977), Hardin (1968; 1977) and Heilbroner (1980).[2] The eco-authoritarian implication of the link between scarcity and political arrangements has been forcefully made by Ophuls. He begins from the assumption that:

> The institution of government whether it takes the form of primitive taboo or parliamentary democracy . . . has its origins in the necessity to distribute scarce resources in an orderly fashion. It follows that assumptions about scarcity are absolutely central to any economic or political doctrine and that the relative scarcity or abundance of goods has a substantial and direct impact on the character of political, social and economic institutions. (1977: 8)

Calling the affluence experienced by Western societies over the last 200 years or so 'abnormal', a material condition which has grounded individual liberty, democracy and stability (1977: 12), he concludes that with the advent of the ecological crisis, interpreted as a return to scarcity (following the 'limits to growth' thesis), 'the golden age of individualism, liberty and democracy is all but over. In many important respects we shall be obliged to return to something resembling the premodern closed polity' (1977: 145). He interprets this, as discussed below, in terms of a (benign) technocratic and theocratic dictatorship. The justification of his anti-democratic stance is basically the traditional argument of 'the ship of state' requiring the best pilots, and the dangers of 'rule by the ignorant' when faced with such a complex and complicated issue as social–environmental dilemmas.

Now while it is perhaps true that democracy does require some degree of material affluence, it is a completely different issue to argue that a diminution in general material prosperity heralds the end of democracy and all its fruits. Additionally, for Ophuls politics and government are understood primarily in terms of 'social survival' rather than other social goals such as 'progress' or flourishing. As such his understanding of politics is as administration or 'zoo-keeping' in Barber's (1984) memorable term. To this extent Ophuls has not established his argument, namely, that a return to material scarcity implies that societies which have previously enjoyed democratic institutions and practices must necessarily revert to authoritarian political forms. It is perhaps more accurate to say that what Ophuls, and the eco-authoritarian position in general, prescribe in terms of substantive political theory is premised and influenced by (a) a focus on the *pace* of social change towards securing (b) the over-riding social goal of *survival*. Hence its apocalyptic, 'doom and gloom' character. It is only by conceiving of the ecological crisis as a 'total crisis' of contemporary societies, in a manner analogous to that described as characteristic of deep ecology earlier, that these theorists can construe ecological problems as a matter of social survival. It is easy to see how from this perspective democratic norms and institutions are superfluous. The anti-democratic basis of eco-authoritarianism lies in its questionable analysis of the nature of the ecological crisis, which emphasizes rapid and anti-democratic social disruption in the name of the over-riding goal of humanity's survival, or more commonly the survival of a particular society. In the eco-authoritarian case, democracy is part of the problem, ill-suited to dealing with social problems which demand immediate and widespread social change. Democracy represents the unleashing of desires, of demands that cannot be satisfied and which have to be authoritatively restrained. Otherwise society as we know it (and perhaps Western civilization) will collapse, as have so many previous societies and civilizations which ignored their dependence upon the environment.

To judge democracy by its effectiveness in securing prosperity is another underpinning assumption of eco-authoritarianism which needs to be criticized. The idea that politics in general or democracy in particular is to be judged to the extent that it makes people happy, for example, either directly or indirectly, is to attribute to democracy something which is not in its gift to deliver. Therefore to assess democracy by these criteria is to commit a category mistake. Democracy as a political decision-making system, after all, is a procedure for making political decisions: that democratic decisions are to make people happy is not to judge the procedure but the product. If democracy is to be judged on the basis of its ability to deliver material prosperity, it is easy to see how this can become an instrumental argument for democracy. If it is material affluence and the social and political stability which such affluence supposedly brings that is valued, it is always possible that a non-democratic form of political decision-making may be superior to democratic forms. Recalling the distinction between liberty and welfare arguments in the previous chapter, eco-authoritarianism is clearly premised on prioritizing basic welfare, i.e. 'social survival', over non-welfare considerations. The eco-authoritarian argument turns on the prioritization of 'survival' and 'security' over ecologically unsustainable and crisis-producing material affluence and the democratic political arrangements that affluence sustains. If the link between affluence and democracy was shown to be mistaken, or demonstrated to be weaker than assumed by the eco-authoritarians, one of the most serious, and common, anti-democratic arguments against green political theory would be, if not undermined, at least substantially reduced.

This is precisely the argument outlined in the previous chapter which, in drawing a distinction between 'welfare' and 'liberty' concerns, presented green arguments for social and economic changes, where any sacrifices or trade-offs were to be limited to those of welfare not liberty. Indeed, it was suggested that decreases in material affluence may be compensated by an increase in liberty as self-determination. If, following standard deontological arguments for democracy, we hold that democracy and liberty are mutually related, then the green argument for socio-economic changes which would have as a by-product a diminution in material standards of living is not inherently anti-democratic. In other words, the green critique of affluence rejects the assumption made by Tocqueville and the eco-authoritarians that there is any necessary connection between democracy and material prosperity. It is economic *security* rather than *affluence* which is important for democracy.[3]

An example of this is Brubaker who, in discussing property relations and environmental management regimes, states:

> A number of countries have assigned property rights to inland and ocean fisheries . . . *Confident that their rights to fish are secure*, fishermen need not waste money building bigger boats and equipping them with more advanced gear in a race to catch fish. (1995: 211, emphasis added)

A result of this would be a more sustainable harvesting of the renewable fish stock, as opposed to the stock itself being 'mined' and overfished. Thus confidence that one can sustain a decent level of well-being can offset attempts to boost appropriation caused by economic insecurity as a result of competition within an 'open-access' situation.

There are at least two ways in which this can be understood. Firstly, against Tocqueville, the green argument may be said to highlight the difference between the relationship of material affluence to the *early* historical development of democracy (the exact subject which Tocqueville was writing about), and the relationship between affluence and the maturing and later development of democracy. In this sense, the green critique of consumerism, for example, can be seen as an argument that beyond a threshold, affluence and the institutional arrangement of society to procure it may begin to 'fetter' democracy and may even hold back further democratization. On this interpretation green anti-consumerism and the arguments of green political economy in the last chapter can be understood as premised on the idea that the further development of democratic institutions and norms may require a less materially affluent society.[4] Thus the relationship between affluence and democracy only holds for instrumental justifications of democracy. These instrumental views are usually utilitarian or economistic in nature, from Bentham and James Mill to contemporary libertarians.

Secondly, against the eco-authoritarians, the green argument for decreased material standards of living may be interpreted as a critique of *liberal* democracy, that is, a particular conception and practice of democracy, rather than democracy *per se*. This is related to the first interpretation. Here the green argument is that the theory and practice of liberal democracy construct a particular balance between 'welfare' and 'liberty' (premised on a particular conception of liberty), which may be inimical to the further development of democracy. In some senses liberal democracy may be more 'liberal' than 'democratic'. It is liberal democracy which fits Tocqueville's argument of the necessary connection between affluence and democracy. Other conceptions of democracy which do not depend upon the high levels of affluence associated with contemporary liberal democracies are of course possible. In other words, if green politics is anti-democratic, it is only anti-democratic in the sense that it criticizes the prevailing liberal democratic conception of democracy. As such green politics can be said to be part of a 'radical' democratic political tradition, a tradition that is not so much 'anti-liberal' as 'post-liberal'. Rather than being anti-democratic, green political theory likes to claim that it constitutes an alternative democratic theory and practice, one which, while critical of liberalism, also builds on some of its core insights and values (Eckersley, 1992a: 30; Doherty, 1996).

A further justification for the contingent relationship between green principles and democratic values and institutions lies in the 'technocratic' dimensions of the eco-authoritarian position. This is most obvious in the authoritative knowledge claims about the scope, severity and components of the ecological crisis which underpin much of the eco-authoritarian view. Invoking the authority of science, particularly ecological science, eco-authoritarianism claims to offer an 'objective', i.e. scientific and impartial, diagnosis of the ecological crisis and an equally 'objective' solution. A useful analogy here is to understand eco-authoritarians as doctors offering an objective, non-negotiable and authoritative assessment of the 'ecological health' of society. Since the ecological crisis is presented in terms of expert knowledge, the implication is that democratic norms are not appropriate to deal with the diagnosis and prescriptions for the ecological health of society. As Saward notes, 'All principled arguments favouring perpetual government of the many by the few are arguments from superior knowledge' (1996: 80). In terms of the social–environmental metabolism discussed in Chapter 5, the eco-authoritarian view is that this metabolism is a 'technical' matter to be dealt with by experts, i.e. those with the appropriate knowledge.[5] In other words, the ecological crisis, like the health of an individual's metabolism, is not a matter in which democratic decision-making is appropriate. As a matter of expert knowledge, not lay judgement, democratic forms of decision-making may be counter-productive to the ecological health of society. The resolution of the ecological crisis, according to Ophuls, requires that:

> the steady-state society will not only be more authoritarian and less democratic than the industrial societies of today . . . but it will also in all likelihood be much more oligarchic as well, with only those possessing the ecological and other competencies necessary to make prudent decisions allowed full participation in the political process. (1977: 163)[6]

On the face of it there seems to be little problem, democratically speaking, with assigning experts to deal with the environment as a technical problem. The point is that the overall character of the 'problem' must be ascertained before its 'technical' aspects can be identified. That is, the distinction between 'technical' and 'non-technical' dimensions cannot be made technically, but needs to be determined politically, by the demos and its institutions. It is practical not instrumental reason which should determine the framework within which the latter can contribute to resolving or coping with social–environmental problems. The decision on the delineation of those aspects of social–environmental problems which are to be made by the few (requiring expertise) and those aspects to be decided by the many (requiring judgement) ought to be made by the many. Depending on how you define the problem, different solutions will suggest themselves, as was clear from the deep ecology view of the 'ecological crisis' as a 'total crisis'. If the only tool one has is a hammer, all problems look like nails, as it were. In the same way viewing environmental problems as 'technical' or 'economic' leads to 'technical' or 'economic' solutions. But if, as suggested in the last chapter, standard economic approaches are flawed when it comes to providing information upon which to base environmental decisions, and leaves the determination of the issue in the hands of a few, then technical expertise will identify the 'problem' as well as the 'solution'. Ecological problems are too important to be left to economists and scientists alone. Identifying a problem as 'technical' leaves decision-making in the hands of the few rather than the many. This is not to deny the absolutely key role to be played by technical expertise in the resolution of social–environmental problems, but rather to stress that such expert knowledge *alone* should not define the problem or possible solutions, or more worrying still, determine the 'language' through which such issues are to be expressed.

The complex of issues involved in deciding the best way to attain an ecologically rational metabolism with the environment are such that making decisions about environmental policy will affect more lives (private and collective, present and future), and to a greater extent than, arguably, any other policy area. The main reason for this is that the economy-ecology material metabolism is the most important aspect of social–environmental interaction, upon which human flourishing depends. Thus even if we allow that one can view the social–environmental problems as 'technical' ones, and that there is a body of knowledge which, and experts who, can resolve them, this does not 'prove' that experts alone should deal with these problems. Even if a problem is technical, there are good reasons why non-technical considerations ought to prevail politically in respect to environmental issues. Simply by virtue of the scope and impact of policies promoting ecological sustainability, non-technical, democratic considerations ought to apply. Since central aspects of environmental policy-making will affect significant areas of private and collective life, it is only right that the demos as a whole decide these questions. This is not to deny the utility and indeed necessity of expert knowledge. Rather, the argument is that this knowledge should not be used to authoritatively determine, as opposed to inform, either the 'problem' or the 'solution'. Once these major issues have been democratically decided, then technical considerations may be appropriate. Experts ought to be 'on tap, not on top', as it were.

The argument against this knowledge-based anti-democratic position lies in rejecting the assumption that social–environmental problems are primarily technical or scientific matters to be defined, analysed and resolved by experts on behalf of, or in the interests of, the wider society. As Saward (1996) argues, the 'political rightness' of an issue ought to be determined by the views and deliberations of the many as opposed to the few. As was pointed out previously, ecological problems and collective ecological management in response to them are matters of both scientific and ethical consideration. As such, coping with social–environmental dilemmas is not simply about diagnosing the ecological health of society, but requires placing this within the context of the totality of social–environmental relations. Providing this context is the democratic aim of collective ecological management. The technocratic basis of eco-authoritarianism comes down to privileging scientific knowledge over other forms of knowledge and understandings of the ecological crisis. The effect of this privileging is to reduce the regulation and assessment of the social–environmental metabolism to a matter of technical or instrumental rationality. Since the ecological crisis is not just a technical matter but requires both claims of knowledge and ethical judgement, it cannot be reduced to a question of instrumental or technocratic manipulation to be left to experts. If one accepts that knowledge is power, to leave the understanding and regulation of the social–environmental metabolism as the exclusive preserve of one form of expert knowledge may crowd out democratic forms of decision-making. This is of course an old anti-democratic argument which can be traced back to Plato's disparaging view of democracy as 'mob rule' by the ignorant. Whereas democratic decision-making on environmental issues implies that these issues be viewed as part of a political-normative process by the demos as a whole, the eco-authoritarian solution requires viewing the regulation of the metabolism as something to be objectively 'deduced' from scientific principles by the competent. This is not to say that green politics is anti-science, but that scientific or technocratic assessments of social–environmental relations should be placed within the wider political-normative context of those relations. The ecological crisis is not simply a combination of technical problems to be 'solved', but also a set of collective moral dilemmas.

A variation on this theme can be found in those analyses of ecological issues where solutions must partake of some definitive (often metaphysical) 'truth' in respect to social–environmental relations. From such perspectives a 'proper' social–environmental metabolism is derived not from objective scientific principles, as in the technocratic-authoritarian argument above, but from the revealed truth of some metaphysical or spiritual system of belief. As outlined earlier, this analysis of the ecological crisis in terms of truth can be seen in the deep ecology position. Whereas the technocratic anti-democratic view prioritized the technical aspects of social–environmental relations over their normative context, deep ecology views the normative context in religious terms, and exhalts that above everything else. While not wishing to claim that the views of deep ecology are inherently anti-democratic, there is a potential tension between its spiritual/metaphysical view of the ecological crisis and its commitment to democratic forms of decision-making. If the ecological crisis is a religious crisis, then the search for a solution must be a search for the 'true' solution which can only be deduced by revelation. In a manner similar to technocratic solutions to social–environmental relations, spiritual perspectives also see the resolution of ecological problems in terms of 'discovery' rather than 'creation'. It is not surprising to find that technocratic and spiritual views are often fused together in eco-authoritarian arguments. A good example of this is Ophuls's argument for a 'priesthood of responsible technologists' (1977: 159) to take charge of regulating the ecological health of society. A similar position has been recently articulated by Mills who states:

> Personally, I would not have much objection to surrendering some, or perhaps most, of my democratic rights to a loving and trustworthy green council which would pursue ends with which I agree and which would benefit me and my family. It is quite on the cards that such sacrifices may be necessary at a practical level and we will have to risk the abuse of the democratic process. (1996: 113)

From a deep ecological viewpoint, Devall's call for spiritually enthused 'eco-warriors' (1988: 196–302) to take care of degraded landscapes is also close to the fusion of spiritual and technocratic arguments which can underwrite non-democratic forms of resolving social–environmental problems. What both the spiritual and technocratic non-democratic perspectives share is a view of the ecological crisis in terms of a particular form of 'objective' knowledge. In the spiritual case the resolution of the ecological crisis as an issue of 'faith' is not open to democratic deliberation.[7]

The general point to be taken from this overview of eco-authoritarianism is that there is a potential tension between green politics and democracy if there is a constitutive relationship between democracy and material affluence or if the ecological crisis is viewed primarily as a matter of either 'survival' or 'salvation'. In both cases democratic forms of decision-making are superfluous, counter-productive or in some way inappropriate to dealing with problems within the social–environmental metabolism. The next section discusses the issue of the role of science in democratic environmental decision-making in more detail.

Science, Knowledge and Democracy

To adequately address the technological and scientific roots of the ecological crisis demands that democratic norms and institutions be extended to what Beck calls 'techno-economic sub-politics' (1992: 229). For Beck the resolution as opposed to the displacement of ecological problems calls for a proactive, *ex ante* perspective, as outlined in Chapter 5. He interprets this *ex ante* position to imply the democratic regulation of technological development. As he puts it, 'The demand is that the consequences and organizational freedom of action of microelectronics or genetic technology belong in parliament *before* the fundamental decisions on their application have been taken' (1992: 229). This argument does not imply either a rejection of science and technology or that they are the sole causes of the ecological crisis, as some early green critics such as Commoner (1971) argued. Rather, in a manner analogous to Gorz's (1989: 111–27) claim that the ecological crisis requires the democratic limitation of economic rationality, Beck's analysis calls for the democratic regulation of scientific and technological practice. In other words, one may interpret the green suspicion of technology and scientific knowledge as motivated from democratic concerns about expert-centred forms of decision-making, notably their secrecy, non-accountability and centralized character. Central aspects of the green critique are thus not anti-science in the manner some observers have maintained (Yearley, 1991; O'Neill, 1993: 148–55). Rather, the green critique of science and expert forms of knowledge and practices ought to be interpreted as claiming that expert knowledge is a necessary but not a sufficient condition for environmental decision-making (O'Neill, 1993: 147). Effective collective decision-making with regard to the determination and regulation of a social–environmental metabolism requires both expert knowledge and lay judgement. This point is further developed below, in terms of the lay and expert composition of the deliberative 'decision-recommending' bodies increasingly used to help make environmental policy (Jacobs, 1997). It was also explicit in the claim that one of the distinctions between ecological modernization and collective ecological management was the greater opportunities for democratic involvement in environmental decision-making, and democratic accountability, in the latter relative to the former.

The imputed anti-scientific outlook of green political theory which is often used as evidence of its regressive, anti-modern stance is thus more apparent than real. While greens may be suspicious of an exclusive reliance on scientific knowledge on the basis that such forms of decision-making can lead to non-democratic results, a positive appreciation of science and technology is essential to the green position. Just as there are democratic reasons which can be advanced for the green critique of economic growth, likewise there are democratic considerations in the green critique of science and technology.

The question to be addressed is the place of science within a democratic society and the place of science as part of the democratic political process of collective ecological management. Whereas techno-optimistic arguments are largely prefaced on the assumption that experts will find solutions to ecological problems with little or no input from the non-expert population, the incorporation of science within green politics assumes that the application of science is *within* rather than *beyond* democratic regulation. This has to do with the fact that scientific knowledge and its technological application can have effects on individuals and that those affected ought to have some say in how science is used. On this account the green democratic argument is for people to have more say in more and more areas of their lives. Within the context of contemporary societies Beck (1992) uses the revealing metaphor of the 'experimental society' to describe how an unregulated, unaccountable technological and scientific establishment turns society into a laboratory without the consent of, and often unbeknown to, the individuals its 'experiments' affect.[8]

Democracy can also be defended as the most appropriate collective decision-making procedure under conditions of uncertainty. Given the often high levels of uncertainty and risk that social–environmental interactions display it would seem that ecological rationality requires that institutions regulating these interactions be as self-reflexive and open-ended as possible. As Hayward points out:

> Given the likelihood of uncertainty and disagreement about knowledge of means, let alone ends, it would therefore seem important that a degree of democracy be allowed into scientific processes in general; and this point would apply *a fortiori* to processes of policy-making. (1995: 186)

While having doubts about the desirability of insisting on democratic norms *within* the production of scientific knowledge itself, I endorse Hayward's point that democracy can be justified on the grounds of uncertainty when such knowledge is used in policy-making. In contrast to the eco-authoritarian argument based on expert knowledge and certainty, general uncertainty and disagreement about the causes, extent and possible remedies for social–environmental problems underwrite the necessity for democratic, open-ended decision-making procedures.

One aspect of this is the necessity for public debate and discussion based on the fullest information available, for it is only by such debate that possible solutions will emerge, since there is no one expert system of knowledge which can be guaranteed to yield an ecologically rational social–environmental material metabolism.[9]

The argument for a democratic as opposed to a non- or anti-democratic context for deciding the equilibrium social–environmental metabolism can be readily seen when one thinks of the disagreements within science itself about key environmental issues such as global warming, ozone depletion, biodiversity loss, pollution, energy and resource depletion. Only a democratic, 'open society' can hope to make good (as opposed to 'true') decisions regarding the material interaction between society and its environment which can command widespread support. Hayward's suggestion for a 'degree of democracy' within science itself is perhaps best understood as an argument for the desirability of the free flow of information and informed debate within science, and not that scientific judgements are to be made on the basis of majority rule. However, the necessity for democratic norms when science is used in policy-making are clear. As Beck has argued, 'Only when medicine opposes medicine, nuclear physics opposes nuclear physics . . . can the future that is being brewed up in the test-tube become intelligible and evaluable for the outside world' (1992: 234). For Beck, following Popper to some extent, democratic decision-making is a form of institutionalized self-criticism which he sees as 'the *only way* that the mistakes that would sooner or later destroy our world can be detected in advance' (1992: 234, emphasis in original). Hence the green stress on freedom of information to permit citizens to make informed choices, and also to combat the anti-democratic possibilities of 'scientism', i.e. science as ideology. This is discussed later in terms of the necessity for a new science–policy relationship or culture.

This argument for democracy *qua* decision-making under uncertainty greatly alters Saward's (1993) suggestion that it is the imperative, end-orientated nature of green politics which implies an instrumental attachment to democracy. One of the central objectives in rethinking green politics is to see that it is not geared towards the discovery of some scientific or metaphysical *truth* regarding social–environmental relations, but rather is concerned with the creation of agreement in respect to those relations. At the same time, there is no 'final solution' to social–environmental problems; rather there are coping strategies. The connection between the experimental nature of social–environmental harmony and democracy also serves to underwrite a non-monistic view of ecological management. According to Norton:

> Since we do not yet know what activities are consistent with protecting the complexity and energy flow in natural systems, and since these will vary from system to system, we do not have a single 'ideal' to guide

> management . . . Conservation biology becomes a part of a social experiment, a part of the search for a viable concept of the good life for human inhabitants of the landscape. (1991: 147–8)[10]

Ecological management must be rooted in a democratic, open and free society in which new forms of social–environmental adaptation, knowledge, can be developed. Thus Beck's argument for a democratic 'self-reflexive' society may be seen as an ecologically updated version of Popper's 'open society'. Democracy, therefore, is accepted on principle as the best procedure for making decisions under conditions of often radical uncertainty. In other words, this conception of green politics does 'embrace uncertainty and . . . the need for constant self-interrogation' (Saward, 1993: 77) as non-negotiable principles.[11]

This need for self-interrogation which underpins democratic decision-making is at the heart of Beck's ecological argument for 'reflexive modernization' and a reconceptualization of social progress. In Beck's analysis, 'social progress' (1992: 203) ought to be understood as institutionalized self-criticism (reflexivity). A major aspect of this involves increasing opportunities for citizens to deliberate and recalibrate the regulative *principles of modernization themselves*, and not just specific policies associated with achieving some 'given' conception of modernization. In other words, green arguments for democracy can be understood as arguments for a different type of social progress, as indeed arguments for sustainability (including sustainable development) are, in part, arguments for a different type of society. This can be readily seen even within 'reformist' conceptions of 'sustainable development', where it is explicitly linked to improved democratic arrangements and the promotion of basic human rights. This is another example of how in practice green politics is not just compatible with but actively promotes democratic forms of decision-making. The tension between modernization and democratization outlined below can, from a green perspective, be lessened if modernization itself is partly understood as a process of democratization as opposed to being mainly viewed in terms of increased individual material wealth. A good example of this revised understanding of social progress concerns the green suspicion of technologically led economic growth which is largely beyond democratic control but which has great and far-reaching effects on the citizens of the demos and indeed various classes of non-citizens (both human and non-human). A green commitment to democracy can be understood as expressing a desire that technological and scientific development be subject to the ultimate authority of the demos. Although the search for the truth is a good, it is not the only or the highest good. As O'Neill puts it, 'A proper understanding of the value of scientific knowledge involves the acknowledgement of limits in the means to and objects of knowledge' (1993: 165).

An obvious example that green politics is concerned with matters of science as well as political and ethical issues is found in the ideas of sustainability and sustainable development. Both sustainability and sustainable development share a fundamental characteristic of having normative and scientific dimensions. That sustainability is a normative concept should be obvious. It embodies a particular moral attitude to the future, expressing for example how much the present generation care for and are willing to make sacrifices for descendants and how, and to what degree, non-humans figure in this process. Making sustainability a co-ordinating social value and practice cannot be left up to 'specialists', since it is not simply a matter of expertise but one that requires ethical consideration. Sustainability calls for judgement rather than uncovering any *ex cathedra* 'true' account of social–environmental relations. Arguments for sustainability usually propose wide-ranging changes in the present organization of society, particularly at the economy–ecology level, in the name of those yet to be born. The consequences of realizing sustainability in social practices are so widespread, and the issues raised so important, that it deserves democratic rather than non-democratic articulation, as indicated in the last section. Even if there is agreement on a general outline of sustainability that ought, for moral as well as prudential reasons, to be socially instantiated, the fleshing-out process has only begun. For a start, as it stands it is far too abstract, being silent on many things. How far in the future must we look? One, three or 50 generations hence? What are we to pass on? What sacrifices are ruled out?

Such questions cannot be answered purely scientifically or metaphysically (that is objectively given), but because of their normative content can only be articulated politically (that is, intersubjectively created). And for traditional reasons we can say that this political process ought to be a democratic one. In one sense greens can ask why they should find new grounds for their adherence to democracy that are different from those advanced by socialists or liberals. The indeterminacies thrown up by sustainability require political adjudication, and given that the policies flowing from any conception of sustainability are likely to have a widespread social impact, leaving few citizens' lives untouched, it is uncontroversial to hold that they should have some say in its articulation and formulation. That is to say the indeterminacy of the principle calls for citizen deliberation, while its translation into policies and laws calls for their consent and, equally important, their participation in achieving it.

Democratic Institutions and Democratic Society

While democracy can be understood in terms of certain types of institutions, such as representative government, and certain institutional arrangements, such as the division of powers and constitutional checks and balances, this focus on the institutional or procedural understanding of democracy does not capture the full normative force of what democracy is. As Macpherson notes, 'Democracy is to be understood as a quality pervading the whole of common life . . . a kind of society' (1973: 15–16). This is also one of the understandings of democracy in Tocqueville's thought. According to Holmes, for Tocqueville democracy specifies 'on the one hand, a social arrangement and, on the other hand, a political system' (1995: 23). Democracy can thus be viewed as a society of equals, a type of social life in which there is an absence of legally maintained class or caste hierarchies: a political decision-making procedure as well as a type of society. The two of course are related. It would be difficult to establish and maintain a democratic society without democratic political institutions. Examples of accounts of democracy which embrace democracy as both a political and a social system include the classical liberal writings of Mill, as well as the more radical democratic tradition of Rousseau, Paine and Marx. One implication of democracy as a society is that it calls for virtues and values such as independence, openness, tolerance, reasonableness and equality to be spread over a wider range of issues and spheres of social life than is required within democracy viewed purely as an institutionalized decision-making procedure. As will be argued below, the link between green politics and democracy cannot be a purely institutional matter. Green democratic theory is thereby concerned with the creation of a 'democratic society' and culture and not just a more democratic political system.

Evidence for this can be found in the green political strategy of 'marching through the institutions', where a green party taking over the reins of the state is seen as a necessary but not sufficient condition for creating a more democratic or green society. As Doherty puts it, 'The greens have responded to new conditions and issues with a distinctively modern strategy based on accepting the limits of the state in guaranteeing social and political change' (1992: 102). However, as indicated in Chapter 5, a green view would also voice concerns over the desirability, and not just the possibility, of overly state-centred social and political change.

Offe and Preuss (1991) bring out this distinction between democracy as an institutional arrangement and as a conception of society in their discussion of the distinction between the American and French revolutions. The American democratic model was a classic example of what they refer to as a 'political revolution', while the French revolution and its model of democracy was rooted in a more fundamental 'social revolution'. The American democratic model is the template 'liberal democratic model' in the sense that it

> relieved the sovereign people from the heavy burden of a nearly sacred task to define and implement the common good. Instead, the model

> restricted itself to the task of devising institutions (such as the natural right to private property and the division of powers) which (a) allowed the individuals to pursue their diverse interests and their particular notions of happiness, thereby at the same time (b) avoiding the danger of an omnipotent government imposing its notion of collective happiness upon the people. (1991: 149)

This model of democracy, they point out, echoing other critical analyses of liberal democracy (Pateman, 1970; 1985; Macpherson, 1973; Barber, 1984), makes little demand upon individuals. Unlike the French or 'republican' version of democracy, the liberal model is argued not to 'enable citizens to be "good" citizens, i.e. citizens committed to the common good' (Offe and Preuss, 1991: 153), but rather to enable them to fulfil their individual interests. Democratic politics conceived instrumentally as a method for the pursuit of individual or group self-interest, cashed out (to use an appropriate term) as material consumption or affluence, may represent a danger to democracy. To some extent, democratic politics is lessened when it becomes overwhelmingly concerned with 'managing the economy', and the securing of ever increasing levels of economic growth, to be privately consumed.

The problems with an overly constitutional approach have been recognized by Kymlicka and Norman who argue that, 'it has become clear that procedural-institutional mechanisms to balance self-interest are not enough, and that some level of civic virtue and public-spiritedness is required' (1994: 360). To paraphrase Clausewitz, modern democratic politics as economics by other means may be bad for the democratic health of society. In this sense the recent green concern with consumerism has roots in older debates around the dangers of commercial society to political democracy. Just as Mill and Tocqueville saw the dangers of commercial society for democracy in terms of its individualism and the elevation of private concerns over public ones, contemporary greens often inveigh against 'consumer society', for similar reasons. If the resolution of environmental problems requires a (moderate to strong) sense of collective purpose as green politics suggests (Achterberg, 1996), then *liberal* democracy may fall short of producing the requisite sense of collective purpose. This is a moot point, and although not a direct concern here, the critique of ecological modernization can be seen as partly stemming from a sense that it is too tied to an institutional democratic approach to social–environmental issues. The ecological deficits of liberal democracy do not underwrite the ecological superiority of anti-democratic approaches, as indicated above, or non-democratic, economic approaches, as argued in the last chapter. In other words, democracy is not condemned on ecological grounds simply because of liberal democracy's imputed ecological failings.[12]

Green politics as collective ecological management rests on the 'greening' of the democratic culture of liberal democratic societies. Thus collective ecological management can be seen as a development from and complement to ecological modernization which works at the level of 'greening' the nation-state and the formal economy. This 'greening' of the democratic culture implies tapping into the energy of a democratic society and using it in the service of ecological goals. This energy stems in part from the free and open nature of democratic society, discussed below in terms of 'the spirit of association' and experimentation. As Paehlke notes, 'Democracy, participation, and open administration carry not only a danger of division and conflict, but as well perhaps the best means of mobilizing educated and prosperous populations in difficult times' (1988: 294–5). In this way green politics seeks to transform not just the institutional structure of presently existing liberal democracies but also the wider cultural context within which those institutions are situated and which they also help create and sustain. One of the defining aims of green politics is that it sees its 'political project' as involving wider cultural transformation. What is suggested here is that green democratic theory and practice are a central location from where this wider cultural transformation begins. In other words, the green focus on democracy has to do with the 'greening' of the existing democratic political culture as the starting point for a wider cultural transformation that green politics ultimately desires and requires. Institutional restructuring from a green point of view must be understood as part of a deeper process of cultural transformation. At the same time political democracy provides a procedure within which cultural contradictions can be publicly raised, coped with and possibly resolved.

Modernization and Democracy

For many greens, economic modernization, whether in state or market form, has historically been associated with non- or anti-democratic political results. According to one version of this argument, modernization is premised on the enclosure of the ecological commons in their various forms (Goldsmith et al., 1992; Wall, 1994). The green argument is that economic modernization, the industrialization of society and economy, which is typically equated with 'progress', is not unambiguously a positive development. There are serious costs which are under- or misrepresented, and/or unjustly distributed. This neglect of the various social and ecological costs of industrial progress is the key to understanding the green critique of contemporary industrial societies.[13] The analysis of progress also serves as a guide to the green attitude toward 'modernity'. If we understand modernity in terms of its industrial and democratic components, then the green democratic position can be viewed as suggesting that there can be a contradiction between the industrial modernization of society and democratic modernization.

A less materially orientated politics does seem to call for a major change in the role of the state. According to Weber (1968), there is a positive correlation between 'modernization' of the socio-economic system, which calls for an increasing complex social division of labour, and the growth of the state and an attendant increase in political centralization and bureaucratization. That is, economic modernization has historically been associated with *both* market and state institutions increasingly regulating social affairs in general (Polanyi, 1957; Tilly, 1992) and social–environmental affairs in particular (Walker, 1989). The analysis outlined in the last chapter, where modernization was portrayed in terms of the 'disembedding' of economic relations from social relations, is consistent with the Weberian thesis. What these perspectives on industrial modernization point to is the dominant position of the state in relation to civil society within the context of a 'modernizing' society. Indeed, the processes of state- and nation-building are traditionally seen as constitutive parts of a modernizing society.

Clearly the opposite seems to hold: that is, 'demodernization' or 'remodernization' may suggest a reduced role for the state and its bureaucracies in economic and social life. As Doherty puts it, 'Once the faith in technological and unlimited growth has been challenged, the need for state-directed maximisation of productive potential falls away' (1992: 101). Equally, as suggested earlier in the critique of free market environmentalism, this would also diminish the need for or justification of free market maximization of production and consumption.[14] This line of argument is of course related to the argument in the last chapter concerning increased economic self-reliance as a central principle for the reorganization of the economy. The more people do things for themselves, the less there is need for extensive state involvement or extensive market relations. It is important to point out that this need not lead to a rejection of the state or the market. As was suggested in Chapter 4, the conception of green politics being defended here does not aim to create transparent social relations. So long as macro institutions such as the nation-state and impersonal markets exist, social relations and social–environmental relations cannot be transparent. All that is being suggested here is that the green critique of economic modernization can be understood and defended on democratic grounds. The latter has to do with the redefinition of the division and relationship between state and civil society, consistent with the role of the state within collective ecological management.

A central element of this has to do with the reconstitution of the state as an enabling state within the context of this reformed relationship, and the distinction between capitalist economic organization and a local market economy as indicated in the last chapter. A redefinition of these relationships is a necessary but not sufficient condition for green democratic arguments. This redefinition partly requires a 'simplification' of economic life which would enhance the prospects for democratic decision-making and democratic norms throughout society as a whole. The less complex social life becomes the less there is a need for large-scale organizations such as current state agencies and economic units. At the same time a less complex social life is consistent with a material basis which is not as damaging and extensive as one which supports a complex one. While this should not be taken as an argument for rendering the whole of social life transparent through simplification, the creation of a less complex web of social relations has the potential to contribute to enhanced democratic practice through decentralization and the development of more 'human-scale' organizations. This calls for a rearrangement of the relationship between state and civil society and within civil society rather than a return to the *gemeinschaftliche* vision criticized earlier. Social complexity cannot be eliminated completely without extensive social restructuring which would herald the break-up of contemporary national societies as they presently exist. The decentralization rather than the deconstruction of nation-states is more in keeping with the green political project of collective ecological management.

An additional factor is that although an anarchistic arrangement may be ecologically rational in the sense of improving the sustainability of social–environmental material exchanges at some established metabolic rate, it is less likely that the establishment of this same social arrangement would be inherently superior to the presently existing system of nation-states in effecting ameliorative and restorative environmental changes. This is because many environmental problems are large-scale and interrelated and cannot be decomposed into smaller components (Dryzek, 1987; Goodin, 1992; Martell, 1994). Now while this issue of scale and complexity does not of itself indicate that large-scale arrangements necessarily require anti-democratic forms of collective decision-making as suggested by eco-authoritarianism, the question of scale and complexity does indicate that the form democracy takes will be different. This is related to the status of direct democracy within green political theory, which suggests the ecological advantages of representative democratic institutions. The net effect of an eco-anarchist deconstruction of the system of nation-states and the creation of small-scale, bioregional units is that the democratic 'rule of the people' with regard to social–environmental interaction may not be improved. The democratic character of green political theory may in fact be better preserved by retaining, but transforming, the nation-state as the basic 'management unit' rather than opting for a multiplicity of autonomous bioregional units. As was suggested in Chapter 5, the democratization of the nation-state may support green democratic arguments better than its abolition.[15]

That a democratic political system has no necessary connection with ever increasing levels of material consumption is a touchstone of green democratic arguments. More important to a democratic polity is a well-developed 'democratic culture', a shared sense of citizenship, plurality and socio-economic and political equality. Plurality and equality are more significant than prosperity as preconditions for an ongoing and vibrant democracy. The advent of scarcity and limits is taken as an opportunity by green theory to redefine basic political concepts. It asks us to consider the possibility that human freedom and the good polity do not depend, in any fundamental sense, on increasing levels of material affluence. Indeed, there may be a trade-off between democracy and material well-being.[16] Beyond a certain threshold, greater increases in the latter may be accompanied by decreases in the former. This is the logic of the prescient quote from Tocqueville earlier. A less materially affluent lifestyle may be consistent with enhanced democratic practice since the decrease in complexity, social division of labour, inequality and hierarchy allows the possibility of greater participation by individuals in the decisions that affect their lives and those of their communities. For example, a shift away from economic growth as a central social goal would undermine the justification of socio-economic inequalities on the grounds that they are necessary 'incentives' to achieve economic growth. At the same time, as early proponents of the steady-state economy pointed out, the shift from a society geared towards economic growth to a society where material growth is not a priority may lead to more extensive redistributive measures (Daly, 1973a). This redistributive aspect to the green critique of excessive material development echoes the socialist critique of the disparity between formal political equality and socio-economic inequality within capitalism.

Unlike the socialist critique the green argument is against economic growth *per se* rather than simply against the capitalist organization of an economy geared towards growth. One of the reasons for this is the green argument that economic growth and modernization demand the creation of organizations, institutions and forms of social relations which may be inimical to democratic practice. Much of the green critique of contemporary society has to do with the contention that large-scale organization and non-democratic forms of social relations are part and parcel of the modernization process. The modernizing imperative, whether it takes a socialist or a capitalist form, has non- and potentially anti-democratic costs. Large-scale forms of production (which require centralized, hierarchical, non-democratic organization), mass consumption, individualism and a breakdown in social solidarity as effects and conditions of modernization can sap the democratic vigour of a society even while maintaining democratic institutions. Modernization in producing a mass society produces a limited and constrained form of democracy in which periodic elections become 'beauty contests' between different political parties, or elites, who are judged on their ability to produce economic growth. This 'protective' view of democracy is of course to be preferred to the authoritarian paths to modernization taken by communist, fascist and other anti-democratic regimes.[17] At the same time protective democracy has its advantages as a procedural conception of democracy. The point, however, is that such protective forms of democracy fall short of creating a more democratic society and enhanced democratic institutions. If the most democratic system economic modernization permits, or is compatible with, is a liberal pluralist system, then those wishing to democratize the state and civil society, such as the green movement, have good reason to see modernization as a fetter. If one accepts that selective demodernization may lead to popular pressure for more egalitarian distribution this in turn may strengthen the equality that is at the heart of democracy. In other words, the green democratic argument is that a well-established and vibrant democratic culture can provide a basis for the democratization of political institutions.

To return to the eco-authoritarian argument above, there is no necessary reason to suppose that a less materially affluent society must lead to a non-democratic politics. It is only if the present unequal distribution of socially produced wealth is maintained *in the face of greater pressure for a more egalitarian distribution* that this is so. One reason for greater pressure for a more egalitarian distribution of wealth is that, apart from an acknowledgement of social and ecological limits to growth, in an ecologically modernizing society, wealth production and distribution are *more transparent* than in an economically modernizing society. In societies such as the welfare states of the West, their legitimacy is, in part, on a commitment to lessen inequalities via redistributive measures. At the same time, providing 'environmental quality' or 'environmental security' is also proving to be an increasingly important feature in securing political legitimacy. Now in a non-growing economy (or a less growth-focused one) it is highly unlikely that the majority of citizens will accept as legitimate the unequal distribution of a static economic pie. The justification of an unequal distribution of socially produced wealth cannot be that it is required for procuring greater wealth production. In short, with the shift to a less growth-orientated society, the normative basis for social co-operation needs to be renegotiated. Within an ecologically remodernizing society, 'social progress' is necessarily and explicitly political. As Beck points out, within what he calls 'industrial society', '"Progress" can be understood as legitimate social change without democratic political legitimation' (1992: 214). Within 'post-industrial' society, such as the type of society green politics is concerned with, 'progress' itself may be more open to democratic deliberation in a way which is not possible in a society geared towards economic growth.

While not doubting the social disruption and problems that a drop in material standards of living may bring, social solidarity and order need not be threatened by this so long as the costs are shared equitably throughout the whole society. Thus there is an important connection between arguments for sustainability and social or distributive justice: indeed politically speaking this is likely to be the most important part of any transition to a more sustainable society. In all likelihood it is only if some members of society are forced to accept lower material standards relative to others that severe social disharmony will emerge. As theorists of taxation have noted, it is not so much the imposition of a collective burden on a people that rankles them as the spread of the burden throughout society. At the same time, denying people the opportunity to make fundamental decisions regarding the content as well as distribution of these burdens would also make for social disharmony and a possible authoritarian reaction. Demanding loyalty and compliance without 'voice' is not possible within a democracy, and as suggested later, compliance with laws and regulations will be enhanced if people themselves have some say in shaping and reshaping them. Also crucial here are arguments which present the transition to a more sustainable society not in terms of falling standards of living, but rather as implying a shift to more 'public' forms of wealth. For example, as Jacobs (1996) has pointed out, a more powerful argument for sustainability involves seeing it as compatible with economic development in terms of public investment in infrastructure, education, health, housing etc. That is, economic development compatible with ecological sustainability requires development to be understood and judged less in terms of year on year increases in personal disposable income and more in terms of public investment.

This conception of ecological modernization (unlike that version criticized in Chapter 5), in depending for its success on creating a greater sense of common purpose, will clearly be strengthened by tapping into existing resources of shared identity. The premise of green democrats is the idea that those affected should be considered as the relevant demos. One could say that a green democracy needs to be issue-sensitive and boundary-indifferent. The boundaries refer to the demarcation of the different interests that green democratic theory seeks to include.[18] The challenge suggested by green politics for contemporary democratic nations is whether they have the foresight and courage to make the further extension of democratic norms and procedures a matter of necessity, as much as a matter of desirability, in addressing the ecological dilemmas facing them.

Discursive Democracy

There is a good deal of support in the literature that the view of democracy which best fits with green politics is a communicative or deliberative model (Saward, 1993; Barry, 1996; Dobson, 1996a; Eckersley, 1996a; Jacobs, 1996). Green arguments for democracy can be said to rest partly upon the integrative function of democracy. This refers to the manner in which democratic decision-making allows the (always provisional) determination of the social–environmental metabolism to be affected by arguments drawn from various sources of knowledge, perspectives and groups. Green politics as collective ecological management is about the collective judgement of the totality of relations which constitute a society's metabolism with its environment. Only an open-ended communicative process such as democracy can call forth and possibly integrate the various forms of knowledge that an ecological rational metabolism will require to command widespread support. And the choice of deliberative democracy is that it offers a procedural rule for democratic decision-making, that of *communicative* as opposed to *instrumental* rationality (Dryzek, 1990: 54). In its operation its advocates also claim that it will increase the 'democratic' character of society in general. Here, the choice of deliberative democracy for deciding social–environmental problems is that it will lead to policies which are made up of that combination of *both* communicative (understood as ethical) and instrumental (understood as scientific/technical) rationality that together constitute ecological rationality in the sense defined in Chapter 3. In dealing with social–environmental relations, as suggested in previous chapters, both instrumental and communicative rationality are appropriate. However, as the ethics of use argument demonstrates, it is communicative rationality which has priority and sets the parameters for the operation of instrumental rationality. This was the position earlier where I argued that the technical dimensions of environmental problems requiring expert knowledge were to be determined by non-technical, democratic forms of decision-making.

The key to understanding the place of deliberative democracy within green political theory is that as a form of collective decision-making it stresses the 'community' over the 'market' or the 'state' as the appropriate location for first-order decisions concerning social–environmental relations.[19] That is, as indicated in the last chapter, a political decision-making procedure is often more appropriate than a non-political one, i.e. the market, in collective decisions regarding the regulation of the social–environmental metabolism. At the same time a deliberative democratic decision-making process is to be preferred over the administrative state making social–environmental decisions for society as a whole (Dryzek, 1995). After all, it is social–environmental relations, not state–environmental relations, that are the subject of collective ecological management. Once the major decisions concerning social–environmental relations have been made democratically (via representative and deliberative institutions), then state or market institutions may be used to carry out those decisions. Discursive democracy in this sense attempts to transform the relationship between state, market and community by seeking, as far as possible, to make both the state and the market instrumental to the democratic decisions of the community.

The choice of a discursive form of democracy within green political theory is closely associated with the public goods character of social–environmental issues. As Jacobs notes, environmental goods are not private: 'Forming attitudes to them is therefore a different kind of process from forming attitudes (preferences) towards private goods. It involves reasoning about other people's interests and values (as well as one's own)' (1997: 219). This is at the heart of the green democratic argument, that only a deliberative or discursive political process will reflect the range of human interests and values in respect to social–environmental relations. Deliberative institutions reflect the public goods nature of environmental problems where individuals do, or ought to, think in terms of the 'public good'. Examples of deliberative institutions include citizens' juries, 'round tables', public inquiries, and the Agenda 21 process as described in the last chapter. 'Citizens' juries' are groups of citizens selected to represent the general public rather than a sectional interest, brought together to deliberate on some matter of public concern. Round tables are a Canadian experiment in which the government chooses representatives from interest groups to try and come to some agreement on social or environmental issues, and to make recommendations (Gordon, 1994). Dryzek also lists 'participatory models of planning, right-to-know legislation, public hearings . . . regulatory negotiation and environmental mediation' (1995: 188) as other examples of 'incipient discursive designs'. For him these are approximations of ideal deliberative institutions, because they are not autonomous 'public spheres' of free discourse since they are associated with the state. The point about these institutions in environmental public policy-making is that they are held to be more appropriate as representing the interests and values of the public on environmental public goods or bads.

These institutions are considered as supplements to, rather than substitutes for, existing environmental public policy-making institutions. Firstly, deliberative democratic institutions are compatible with existing liberal democratic institutions. While deliberative democracy does suggest a more participatory form of democratic decision-making, it is compatible with representative government. Secondly, they are confined to environmental policy-making, though most of the advocates of deliberative democracy believe that their widespread use will create an impetus for the democratization of the state and other areas of policy-making, given that environmental concerns transcend administrative boundaries (Dryzek, 1995; Christoff, 1996). Thirdly, in its real-world approximation, there is no reason to suppose that pure communicative rationality is the only or main procedural standard. Jacobs (1997) suggests that within deliberative decision-making there may be certain desired or proscribed outcomes. According to him:

> society can constrain the deliberative process by imposing on it the requirement to fulfil particular end-values; that is, particular broad conceptions of the public good . . . Certain values (such as racism) could be ruled inadmissible; or the deliberative process could be asked to come to a decision in pursuit of particular ends. (1997: 227)

Again this can be taken as evidence of the supplementary role that deliberative democratic institutions can play as part of the range of democratic institutions which constitute a democratic political system. Finally, the deliberative process could be required to decide using a range of decision-making procedures. For some issues, consensus may be appropriate, for others simple majority rule, and for others a two-thirds majority rule may be suitable. As Tännsjö, in a different context, notes:

> In situations where we want to find out what to consider morally right or wrong, or where we want to reach a decision as to what we are to take as a matter of fact, majority democracy may seem appropriate, while in situations where we have to divide up between us (equals) some amount of money or other economic resources, it may seem appropriate to practice unanimous democracy. (1992: 43)

The essence of the deliberative view of democratic decision-making is its communicative character. It is not voting in private booths, but debate and discussion within something approaching a 'public sphere', which marks deliberative democracy. Since according to Jacobs, 'Attitude formation towards public goods is . . . essentially a public not a private activity' (1996: 217), it follows that a public and deliberative procedure is required in order that these attitudes/preferences towards environmental public goods be created. The point is that policy-making based on aggregating privately formed preferences will not be based on the appropriate information. This information can only be created within deliberative not aggregative contexts.

Within a deliberative rather than an aggregative context, participants are more likely to engage in 'public good thinking'. Jacobs (1996: 8–9) suggests three reasons why deliberative institutions are likely to encourage this. The first is that arguments must be put in terms of the public good. Arguments in terms of private or sectional interests are unlikely to produce majority agreement. The second is that deliberative institutions expose participants to a wider range of perspectives than is likely with private contemplation. Here the representativeness of the participants is crucial, and a standard of communicative rationality which allows *all* arguments/points that participants wish to raise to be raised. The third is that the act of deliberation tends to create a community amongst participants. Communication and contact with others under conditions of respect and equality invite participants to a greater sense of mutuality, solidarity and sympathy. One may add that in cases of conflict, a deliberative setting may also foster a sense of toleration and understanding if not agreement. Thus, one should not judge deliberative democracy solely by its capacity to produce agreement (or consensus in Habermas's abstract account of the 'ideal speech situation'). Deliberative institutions may not themselves solve social–environmental problems in the sense of producing agreement on the right course of action, or policy to be implemented. However, their widespread use may create the conditions which result in agreement. At the same time, it is not proposed that deliberative democratic institutions replace existing representative ones. Rather the appeal to deliberative democracy is to supplement existing democratic institutions. The discursive claim is that certain environmental problems, for example, major land-use proposals where 'preservation' conflicts with 'development', lend themselves to a deliberative rather than an aggregative democratic solution. Thus in the case of a public inquiry into a proposed development project, the remit of the inquiry should permit the possibility of suggesting that the development not proceed. In other words, unlike the position described in Chapter 5 where such participatory forms of public involvement are often limited to influencing *how* development proceeds, there ought to be the opportunity to influence the inquiry's findings based on arguments over *whether* it should go ahead.

The advantages of a deliberative approach can be seen by looking at the distinction made between the different environmental attitudes people have as citizens and as consumers. One way of viewing this distinction is Sagoff's (1988) argument that individuals as consumers are guided mainly by considerations of their own interests, whereas as citizens they have to, or ought to, place the latter within the context of a common good, which accommodates the interests of others as well as their own. While Sagoff's position brings out the public/private dilemma within environmental issues, it is based, as Keat (1994) points out, on an overly restricted conception of consumption as a purely private activity. According to Keat:

> consumption . . . is not merely something that we pursue as individuals: it is also a major element of the shared values of the local culture. So when, as citizens, people debate the nature and implications of their conception of the good society, they will find that a central element in that conception itself concerns the value attributed to consumption . . . Hence what Sagoff represents as a tension within individuals between their roles as consumers and citizens might better be seen as a tension within the culture between the values of consumption and of nature – one that they have to address as citizens. (1994: 343–4)

Sagoff seeks to replace consumer interests with those of the citizen, that is, replace economistic valuations of the environment with political-normative ones. However if consumption is itself a social value and not simply a private activity, then, as Keat suggests, the resolution of social–environmental problems requires a wider cultural context. The green democratic argument is that this cultural contradiction can only be resolved democratically. And this is more likely to be achieved when deliberative democratic settings and institutions supplement existing representative ones. The point about deliberative democratic institutions is that they can bring out the intersubjective character of environmental values, and articulate publicly the different forms of human valuing and bring them to bear in the making of social–environmental decisions. The problem with economistic forms of valuing the environment is not that they are necessarily wrong *per se*, or that they are non-moral, but rather they are wrong when they monopolize the debate, 'crowd out' other forms of valuing and human interests in nature, and are standardly used as the primary form of information upon which to make environmental decisions. Green political theory does not seek to abandon instrumental social–environmental interaction. The legitimacy of instrumental social–environmental relations is, it will be recalled, at the heart of the argument in Chapter 3 for an 'ethics of use'. There it was argued that the normative basis of green politics cannot be abstracted from the existential fact of human manipulation and use of the environment. It is the manner of this interaction with which the ethics of use as a central normative aspect of collective ecological management is concerned. Such a public ethic is the desired outcome of the deliberative democratic process reflecting, as accurately as possible, the collectively agreed set of public and binding norms that are to regulate social–environmental affairs. These norms are the outcome of a deliberative process within which questions concerning both ends and means pertaining to them can and ought to be raised.

It is the normative indeterminacy together with the epistemological uncertainty associated with social–environmental interaction that calls for democratic political deliberation. It is important to bear in mind that this indeterminacy and uncertainty applies to both ends and means within the green political project. If green politics is concerned with securing agreement for an ecologically rational social–environmental metabolism, this implies, as indicated above, that green politics cannot guide itself by seeking to discover some 'true' social–environmental metabolism.[20] Any such notion of one 'true path' is both dangerous and potentially undemocratic since it can function as a way to close debate and discussion. While green politics is ultimately presaged on a belief that there can be a rational harmony between human and non-human interests, it is not supposed that there is only one equilibrium pattern, and it will vary from place to place. Rather, as indicated in Chapter 5, there are a (limited) number of such patterns which may secure an ecologically rational social–environmental metabolism. Choosing which pattern and rate can only be legitimate if it is democratic and involves not just questions about means but more fundamentally about the ends of social–environmental relations. In choosing a particular metabolism one is also choosing a particular pattern of social–environmental relations, a certain mode of collective being, and thus a particular type of society. Just as a virtue ethics view of social–environmental relations asks the question 'what sort of person should I become?', the political analogue of this sees the choice of social–environmental metabolism as choosing to live in a different type of society.

Institutions and Principles for Collective Ecological Management

In this section the focus is on assessing the common perception that green democratic theory must be some variation of direct democracy. With the state and citizen playing such central roles, representative forms of democracy are perhaps more central to green concerns than is usually thought. Highlighting the role of representative institutions is also another way of expressing the 'post-liberal' as opposed to the 'anti-liberal' complexion of green democratic theory.

One of the arguments in favour of representative democracy is that, unlike participatory or direct forms, the 'politicization of everyday life' is not one of its goals. The disputes that occur within representative democracy do not share the same intensity as those that may occur in the face-to-face context of 'strong democracy' (Barber, 1984), or the small-scale, decentralized, self-sufficient communities that pepper the green political literature. In such a context it is often difficult to distinguish a fellow citizen's opinions from her as a individual, and while respect should always be shown to the individual independent of her particular views, under direct democratic conditions this important distinction may become blurred or even broken. Maintaining a central place for representative institutions within green democratic theory entails rejecting the attempt to create transparent social relations. It also relates to the defence of the private sphere mentioned in the last chapter.

For example, the private sphere of the family, as the primary site of consumption as well as the site of decisions concerning procreation, is of central significance in influencing social–environmental relations. However, it is not part of the theory of green politics being defended here that the familial sphere be democratized by, for example, the abolition of the family and the creation of communal or state-controlled child-rearing institutions. Nevertheless, given the importance of population in affecting social–environmental relations, it is clear that it may be necessary to regulate population growth. This means that the decision to have children cannot be left to individual choice alone but has to be taken within the context of how it may affect social–environmental relations. However, there is a world of difference between the democratic regulation of private decisions concerning procreation, which is only one constitutive aspect of the family, and the abolition of the family as a social institution.[21]

Another reason for greens to endorse representative democratic institutions has to do with the green concern to give 'voice' to the interests of previously excluded others. Three different classes of affected interests, which do not at present have any direct democratic representation within the decision-making process, have been identified by green theorists and commentators. These are the interests of future generations, affected foreign nationals, and (parts of) the non-human world (Kavka and Warren, 1983; Dobson, 1996b; Goodin, 1996). Although democracy is necessarily *by* the people and *of* the people, it does not necessarily have to be *for* the people, where the 'people' is understood as a human community presently living within a nation-state. On the face of it, if – and of course this is a moot point – but if the political exclusion of the interests of these three classes of non-citizens is held to constitute a defect in democratic practice, then it is clear that representative institutions offer the most defensible and practical way of including them in the democratic process.

The appropriateness of representative over direct democratic forms is most obvious in the cases of the interests of future generations and those of the non-human world. These groups cannot themselves express and publicly defend their interests, the former because they do not exist and the latter because they cannot communicate their interests. It is therefore uncontroversial to suggest that the only sensible form that their inclusion in the democratic process can take is a representative rather than any direct form. It is only in the case of affected foreigners that it is possible for their interests to be directly expressed and brought to bear in any direct democratic decision-making procedure. However it is important to bear in mind that in respect to these three classes of interests it is not true to say that they have no influence on the democratic decision-making process. Even without separate representative measures, the interests of these groups can be brought to bear if there are citizens who incorporate and consider their interests in making decisions and/or are willing to defend these interests publicly in an attempt to persuade fellow citizens to think likewise. To a greater or lesser degree, it is possible that *any* interest can be brought to bear and have an effect on the democratic process. The only stipulation is that there should be some citizens or groups or parties who incorporate these various interests and/or publicly represent them, seeking to persuade their fellow citizens of the propriety or prudence of taking these interests into account. In many respects the argument for the democratic representation of the interests of non-humans for example is a reflection of the failure of these interests to be reflected within the interests of citizens. In other words, the creation of democratic institutions to represent non-human interests arises partly from the lack of 'green citizenship' and a wider ecological culture. If, as a matter of course, citizens took the interests of non-humans into account when making social–environmental decisions there would be less need for these interests to be directly, democratically represented since they would already be incorporated by citizens themselves. That is, within an ecological culture there would perhaps be less recourse to political or legal means of representing the interests of non-humans.

Whatever interests are to be democratically considered, what needs to be remembered is that it is not the interests *per se* of future generations, non-humans and foreigners that is at stake. Rather what is being considered are these interests as *perceived* by democratic actors in the course of democratic decision-making. To suppose otherwise would be to presume an infallibility and omnipotence which cannot be sustained. As indicated earlier, there are limits to human knowledge of the world. Even the most diligent ecologist cannot be said to be infallible in determining what is in the interests of non-humans, or indeed what their interests are. This leads to the question of *who* is best placed to represent the interests of non-humans: deep or shallow ecologists, economists, lay citizens, environmental managers? The democratic point is that no one group of citizens can be assumed to be the 'true representatives' of non-humans. Rather there needs to be a democratic debate about what their interests are and how they weigh against the interests of humans. Such issues are clearly the essence of normative debates once a particular use or development of nature has been sanctioned. That is, they are side-constraints on *how* we use the non-human world, rather than *whether* to use it. The question is not 'why take the interests of non-humans into account?', but rather 'why not?' and 'how?'

To a greater or lesser extent it is true to say that the interests of all three classes are incorporated by citizens in democratic societies. This is particularly so with the interests of future generations. Consistent with a naturalistic moral perspective is the postulate that humans do care about their descendants, particularly proximate ones. In other words, the interests of future generations are already incorporated within the extended interests and considerations of the present generation. If this is the case, it may be that the interests of future generations can be brought to bear on the democratic deliberations of the present one by a procedural rule which makes it mandatory that decisions be made with the interests of the future in mind.[22] And as many writers have sought to demonstrate, making environmental decisions which take the interests of future human generations into account does go a long way in securing the types of policies proposed by those wishing to preserve and protect the non-human world for its own sake (Norton,

1991; de-Shalit, 1995). For Norton this 'convergence thesis' undercuts much of the deep ecology position. He argues that,

> introducing the idea that other species have intrinsic value, that humans should be 'fair' to all other species, provides no operationally recognizable constraints on human behaviour that are not already implicit in the generalized, cross-temporal obligations to protect a healthy, complex, and autonomously functioning system for the benefit of future generations of humans. Deep ecologists, who cluster around the principle that nature has independent value, should therefore not differ from long-sighted anthropocentrists in their policy goals for the protection of biological diversity. (1991: 226–7)

Long-sighted anthropocentrism is of course a key aspect of ecological stewardship, which as argued in the next section forms the core of 'green citizenship'. The point here is that taking the interests of future generations into account will, as Norton suggests, converge with non-anthropocentrists on environmental policy goals. Policy agreement need not depend upon substantive normative agreement. Differences at the level of pure ethics can be compatible with agreement at the level of applied ethics, or in this case, policy goals. Insisting on agreement on the reasons for action can often be counter-productive. In terms of normative underpinning of environmental policy, *contra* deep ecology, consensus on the reasons for action is *not* as important as agreement on the action itself.

Whatever form it takes, the representation of the interests of these classes may act as a side-constraint on policy decisions that the present generation may make. In giving voice to their interests there is no obligation on behalf of a green democracy to promote or positively enhance these interests. That the interests of these excluded groups be equally considered, publicly registered, and thus included in the democratic decision-making process, is enough. Equal consideration of interests does not guarantee that those interests be equally satisfied or protected. As suggested earlier in the critique of eco-authoritarianism, the justification of democracy should not be based on its ability to secure the satisfaction of welfare interests.[23] However, in the case of future generations, viewed as descendants of the present generation and not as future generations of humanity as a whole, it is likely that decisions will attempt to promote and improve their situation. While democratic theory does not demand that their interests be positively secured as opposed to represented, it is likely that citizens will seek to promote and protect the interests of descendants and future citizens.

That these previously excluded interests are given 'voice' in the democratic process is sufficient to satisfy the demands of green democracy. The process of representing the interests of the non-human world, that they count for something, is a necessary aspect of the democratic determination of legitimate 'use' from illegitimate 'abuse' in respect to social interaction with the environment. At the same time the consideration of the interests of others is supplemented by citizens being encouraged to reassess, re-evaluate and perhaps alter their own interests in the light of democratic debate and deliberation. This last point is discussed below. This argument for considering the interests of others in the determination of social–environmental affairs underwrites the previous argument in Chapter 2 that there is no *a priori* disposition in favour of 'preservation' as opposed to 'development'. What green democratic theory requires is that the interests of the non-human world, for instance, are considered, not that they automatically 'trump' those of humans. Going against the grain of much green moral theory, the position defended here is that humans have no presumptive 'right' to development and non-humans have no presumptive 'right' to preservation.

From a green view, representative democracy may be improved by institutional changes to its operation as well as by supplementing it with more institutionalized opportunities for citizen participation. It is clear that Beck's (1992) argument for 'institutionalized self-criticism' will involve changes to the workings of existing representative institutions as well as the creation of new forms of democratic participation. However, it is not the case that the latter requires creating democratic institutions in order to make social relations transparent. While the eco-anarchist position is generally marked by a desire to make social relations transparent to the individual by breaking society into smaller units and then constructing direct forms of democracy, it is fair to say that the retention of state institutions does acknowledge the opaqueness in social relations. The point however is that the institutional arrangements being suggested here, in conjunction with the arguments in the last chapter for self-reliance and reduced complexity, will create the conditions to make existing social arrangements less opaque and more open to democratic scrutiny and accountability.

Additional procedural demands of green politics are that democratic environmental decision-making be regulated by principles of reflexivity, openness and precaution. These can be taken as meaning that when democracy is faced with large-scale environmental decisions, no decision should be taken whose effects cannot be reversed in the future. This is one way in which to understand the idea that the social–environmental problems of democracy cannot be assumed to be solved simply by more democracy. What is required is 'better' democracy, which in part has to do with better informed democratic decision-making procedures. And enduring within the green movement has been the demand for greater openness in environmental decision-making and greater freedom of and access to environmental information (Paehlke, 1988; Beck, 1995). So while it is no part of green democratic theory to make all social relations transparent, a central aspect of it is to make environmental decision-making open to democratic scrutiny. For Beck this is one aspect of what he calls the

> secret elective affinity between the ecologization and democratization of society . . . The long-term policy towards threats should be slowing down, revisability, accountability, and, therefore, the ability for consent as well; that is to say the expansion of democracy into previously walled-off areas of science, technology, and industry . . . What is important is to exploit and develop the superiority of doubt against industrial dogmatism. (1995: 17)

The significance of the precautionary principle lies in what commentators have called its challenge to 'the established scientific method . . . the application of cost–benefit analysis . . . established legal principles and practices such as liability . . . politicians to begin thinking through longer time frames than the next election or economic recession' (O'Riordan and Jordan, 1995: 193). This principle carries with it notions of best practice in environmental management and good husbandry, values which are clearly compatible with ecological stewardship. As a procedural standard, the precautionary principle can be viewed as simply asserting that decision-makers should act to protect the environment in advance of scientific certainty on the issue (1995: 194). As mentioned earlier, prudence and precaution are rational procedural standards to adopt when making decisions under conditions of uncertainty. One way of looking at the application of the precautionary principle to democratic decision-making is to see it as specifying a range of outcomes which are impermissible, namely those that cannot be altered or reversed in the future. Thus applying this principle would require ruling out irreversible environmental changes for example. Not all options are available as potential comparators for environmental decision-making under conditions of uncertainty. In situations of uncertainty the onus of proof is on the risk creators, those who propose development rather than those who oppose it. The point is that democratic environmental decision-making, where uncertainty and ecological vulnerability are high, calls for prudence and self-limitation. In this respect 'ecological rationality' is a bounded form of rationality. This self-limiting aspect of the precautionary principle can be viewed as an additional self-binding character of democracy. It can also be viewed as a rational decision-making procedure for long-term collective interest.

Thus the application of the precautionary principle can be viewed as the institutionalization of the ecological virtue of prudence under uncertain conditions. To relate this to the points made earlier in reference to the authority of science, the precautionary principle works without the assumption that science will or can determine or provide an agreed conceptualization of the environmental issue at hand. It is precisely because environmental problems are disputed (for example, global warming) within the scientific community that the precautionary principle holds that decisions ought to be made in advance of scientific proof (consensus within the scientific community), which may not be forthcoming anyway. The significance of the precautionary principle has already been stressed previously where it was argued to be a key feature of ecological modernization upon which collective ecological management seeks to build.

What the application of this principle indicates is a challenge to the accepted relationship between science and policy-making, or what O'Riordan and Jordan call the 'science–policy culture' (1995: 208). One political implication of the precautionary principle for O'Riordan and Jordan is the necessity for a '"civic science" or the science of open public debate about determining uncertain futures' (1995: 207) as a crucial dimension of a 'greener' science–policy culture. It calls for a more communicative, sensitive relationship between the two, and shifts environmental decision-making away from technical or expert determination based on known 'facts', and towards making public judgements in the face of uncertainty and controversy (Barry, 1996). As Kai Lee argues:

> Managing large ecosystems should rely not merely on science but on civic science; it should be irreducibly public in the way responsibilities are exercised, intrinsically technical and open to learning from errors and profiting from success . . . Civic science is a political activity. (1993: 161)

Green Democracy, Citizenship and Stewardship

One of the questions green politics addresses, and upon which its practical success depends, is expressed in Elster's statement that 'the central concern of politics should be the transformation of preferences rather than their aggregation' (1983: 35). This was raised earlier, where it was suggested that collective ecological management requires that individual environmental preferences alone cannot be taken as the basis for collective ecological decision-making.[24] This has to do with the problem of endogenous preference, the fact that preferences are not 'given' or fixed, but are unstable and malleable. Endogenous preferences highlight the institutional and contextual nature of preference origination and formation. According to Sunstein, 'The phenomenon of endogenous preferences casts doubt on the notion that a democratic government ought to respect private desires and beliefs in all or almost all contexts' (1995: 197). The institutional context within which preferences are formed was at the heart of the argument in the last chapter where green political economy was argued to be part of the tradition of institutional economics. It was also an argument that environmental preferences as preferences for a public good demand a public rather than a private institutional setting. To get the appropriate information as regards people's environmental preferences one needs to pay attention to the context within which preferences are formed. The green emphasis on preference formation, which was argued to be a central concern of the green conception of autonomy in the last chapter, is thus related to preference transformation. Under different institutional conditions, information or rules, preferences will be different. However, the main point is that

> if the rules of allocation have preference shaping effects, it is hard to see how a government might even attempt to take preferences 'as given' in any global sense, or as the basis for social choice. When preferences are a function of legal rules, the rules cannot be justified by references to the preferences. (1995: 202)

One way of looking at the lack of any pre-political or pre-institutional preferences is to say that preference formation is already political, to make it explicit rather than implicit.[25] Environmental preferences, as Jacobs (1996) suggested in the last chapter, are 'political' in this way, unlike preferences for private goods. Part of what this means is that existing environmental preferences cannot be taken as 'given' since they are the product of, for the most part, unecological background conditions. If it is accepted that preferences are in part 'created' by institutions, and if those institutions can be shown to have developed at a time when ecological considerations were neither important nor known, then there is a case to be made that those institutions may not be appropriate in the face of altered environmental conditions. In the same way one can ask whether present liberal democratic institutions are the most appropriate for dealing with social–environmental problems. At the same time, preferences partly created by the latter cannot be taken as justification for environmental policy-making. Are we obliged to satisfy collective decisions that may be based on faulty and mistaken information? If preferences are adapted, why not create a more extensive 'adaptive context' and opportunities for preference formation and transformation, rather than aggregating them?

One implication for this is that the state's adoption of particular institutions through which to implement policy will be important (Jansen and Osland, 1994). If it chooses market instruments such as those suggested by neoclassical environmental economics, it addresses individuals and groups in society as consumers, and the environmental problem is viewed in economic terms. If however the state uses legal instruments, the environmental problem is viewed in terms of right and wrong, permissible and impermissible, rather than costs and benefits.

And individuals are addressed as democratic citizens under the law. According to Jansen and Osland,

> in the case of laws against pollution, the citizens . . . are encouraged to take a stand on the values the imperative is supposed to promote. The citizens may of course act according to the law, and still disagree with the law and the values that law represents . . . But in doing so, they have to reflect on the discrepancy between their private preferences and interests, and the societal values. (1994: 13)

The point is that in choosing an institution for environmental policy-making one chooses a particular way of presenting the problem and addresses individuals in a particular role or identity. The point about political-legal approaches is that they encourage people to act and think as citizens, to see the environmental problem in normative terms and to assume responsibility for one's actions in terms of doing what is right. This attempt to construct the pattern of rights and duties between citizen and state has been made by Weale (1992: 150–1) in reference to the Dutch National Environmental Policy Plan, and one may extend this argument to suggest that adopting a political approach to environmental problems implies, in part, encouraging citizens to become 'ecological stewards', responsible partners, along with the state and other institutions such as economic organizations, for the management of the environment.

The centrality of citizenship to green arguments for democracy comes from the belief that the achievement of sustainability will require more than institutional restructuring of contemporary Western liberal democracies (Achterberg, 1996). Such institutional changes are necessary, but not sufficient, from a green point of view. The green contention is that macro- and micro-level reorganization needs to be supplemented with changes in general values and practices. In short, institutional change must be complemented by wider cultural-level changes. Deliberative forms of democratic decision-making are preferred, because a deliberative conception of citizenship is more likely to result in political cultural change for which greens argue. Citizenship can be understood as a mediating practice which connects the individual and the institutional levels of society, as well as constituting a common identity which links otherwise disparate individuals as members of a shared political entity. Within green democratic theory, citizenship is a practice within which ecologically beneficial virtues such as self-reliance and self-restraint can be learnt and practised. Although green citizenship is politically based, the activities, values and principles it embodies are not confined to the political sphere as conventionally understood. The virtues one would expect to be embodied in this green form of responsible citizenship, as a form of moral character, would be operative in other spheres of human action and roles.

The green claim to a principled as opposed to an instrumental adherence to democracy is that deliberative forms of 'democratic will formation' permit the possibility of sustainable and symbiotic human interests motivating social–environmental relations. In this sense green democratic citizenship may be understood as a form of social learning, with democratic deliberation, in part, as a public form of pedagogy. However unlike the epistemological arguments which underpinned some arguments for the non-democratic determination of social–environmental relations, the pedagogic nature of deliberative democracy is not about the internalization by the populace of some given truth as determined by experts. Rather, the pedagogic effects of deliberative democracy is a process of mutual learning, the bringing together of various forms of knowledge (both expert and vernacular) and arguments (moral and non-moral) before citizens so that their deliberations can be as informed as possible. This view of citizenship as a form of social learning turns on the view of democracy as a communicative process. It is also related to such practices as LETS, discussed in the last chapter, which can be regarded as forms of social learning and adaptation to changed ecological and socio-economic conditions (Barry and Proops, 1996), as well as the ecological restructuring of the state and economy.

It is also related to the claims of the last section where green democracy was identified as extending the range of interests to be included within the democratic process. Part of the green argument for deliberative forms of democracy is that the latter provide perhaps the best way in which citizens can be persuaded to take the interests of affected non-citizens into account. In other words, deliberative citizenship as a practice within which argument, debate and deliberation are central may achieve by persuasion what the direct democratic representation of the interests of excluded classes may not. The working of the deliberative model of democracy within which green arguments could convince sufficient numbers of citizens of the normative rightness or prudence of considering the interests of non-humans, foreigners or future generations may obviate the need for separate institutional representation of their interests. This is the logic of deliberative democracy: rather than the individual being concerned with her own interests, she is encouraged to consider the interests of all those potentially affected by the democratic process. As Goodin suggests, 'It might be empirically more realistic, as well as being morally and politically preferable, to think . . . of democracy as a process in which we all come to internalize the interests of each other and indeed of the larger world around us' (1996: 18). This view of democracy as a process within which we recognize that we are, to a greater or lesser extent, each other's keeper is clearly compatible with the ecological view which holds that the determination of social–environmental relations within one human society has effects which transcend that society, and the species.[26] It is also in keeping with the view that with the technological power at the disposal of the currently existing generation comes responsibility. It is appropriate that the democratic regulation of ecological risk be effected through internalizing the interests of others. The internalization of the interests of others (both fellow citizens and the classes of non-citizens identified above), as well as the transformation of preferences, as a result of democratic deliberation, will be indispensable to the achievement of an ecologically rational metabolism. The need for a deliberative democratic form within which the interests of others may be considered is that it is the interests of these silent others as perceived by fellow citizens that is internalized. It is only by encouraging the presentation of all possible interpretations of the interests of others that citizens may agree on a considered delimitation of what it is that they collectively owe the future, foreigners, non-humans or each other. While it may be going too far to expect the internalization of the interests of others, as Goodin (1996) suggests, the least we can expect from a green democracy is the consideration of the interests of affected others.

The introduction of communicative rationality to the co-ordination of individual action makes it less likely that the collective result will be ecologically irrational. Enhanced democratic institutions which stress citizen participation and deliberation on collective issues are more likely to avoid the prisoner's dilemma in regard to environmental public goods and bads. The famous formulation of the paradigmatic ecological problem as the 'tragedy of the commons' can be criticized for not allowing purposeful communication between individual users of the commons. It simply assumes a prisoner's dilemma scenario with mutually disinterested and non-communicating 'rational individuals'. However, by introducing a communicative dimension, an intersubjective realm is created which permits the co-ordination of individual activity in such a way that the aggregate effect of individual behaviour is not, as in the tragedy scenario, both collectively and individually undesirable. Democracy understood as communication (Dryzek, 1990), together with democratic citizenship as part of a self-reflexive, social learning process (Beck, 1992), provides some evidence that they can deliver enhanced environmental public goods and avoid or limit environmental public bads (Paehlke, 1988). This is because deliberative or discursive democracy is based on preferences, expectations and behaviour being altered as a result of debate and persuasion, and on binding individual behaviour to conform to publicly agreed decisions. Democratic citizenship in short permits the possibility of the voluntary creation and maintenance of an ecologically rational society–nature interaction, informed by moral as well as scientific considerations. This is because *communicative* as well as *instrumental* rationality characterizes ecological rationality.

Citizenship as a practice can also be used to cleave representative and more participatory democratic institutions. An invigorated, active citizenship is possible within a democratic system made up of both representative and participatory institutions. Unlike the demand for direct democracy which does require the transcendence of representative institutions, the demand for deliberative or participatory democracy does not. This has to do with the argument suggested above that green democratic theory and practice do not require transparent social relations. It also has to do with an acceptance of the idea that the problems of democracy (including ecological ones) cannot be solved simply by more democracy.

Citizenship, as viewed by green democratic theory, emphasizes the duty of citizens to take responsibility for their actions and choices – the obligation to 'do one's bit' in the collective enterprise of achieving sustainability. There is thus a notion of 'civic virtue' at the heart of this green conception of citizenship. A part of this notion of civic virtue refers to consideration of the interests of others and an openness to debate and deliberation. This implies that the duties of being a citizen go beyond the formal political realm, including, for example, such activities as recycling waste, ecologically aware consumption and energy conservation. In these cases there are roles for the formal institutions of local and central government, the constitution, the judiciary, as well as more informal institutions of community, the opinions of fellow citizens, to prevent 'free-riding' by individuals and groups. That is to say a green democratic society will also need to encourage a 'sustainability culture', resources for which are already present in the 'bioculture' of contemporary liberal societies, as well as in the emerging 'ecological culture' of those societies, as evidenced in the generational difference between young and old in support of 'green' or 'ecological' causes and attitudes.

The notion of human stewardship in relation to the non-human world which was introduced earlier is central to green citizenship. Recalling the discussion in Chapter 3 about the importance of establishing when legitimate human use of the environment becomes unjustifiable abuse, green citizenship can be viewed as the practice of ecological stewardship. That is, green citizenship is a collective practice determining an 'ethics of use' which expresses a particular understanding of the stewardship ideal.[27] One may say that the 'good' green citizen is one who most approaches the ideal of the 'ecological steward', a central part of which involves considering the interests of fellow citizens, non-humans, foreigners and future generations. For example, it may be in the interests of a particular collection of non-humans that their habitat be preserved while it may be in the interests of citizens that it be developed. To be a good green citizen does not entail an obligation to actively promote the interests of non-humans or others over one's own, but rather to justify and assess one's interests in the light of the interests of others. In practice this implies that a virtue of responsible green citizenship is a willingness to accommodate the interests of others within an expanded conception of the 'ecological common good', a common good within which one's own good is located. When faced with social–environmental problems good ecological citizens are motivated to seek solutions in which human and non-human interests are rendered as compatible as possible. In order to satisfy as many interests as possible of course requires that there be a willingness to compromise as well as an openness to persuasion through public debate.[28] *Ceteris paribus*, the good of satisfying as many interests as possible is a key goal of green politics and ecological stewardship.

However, responsible citizenship, as Held (1991b: 23) points out in reference to socialism and the democratic empowerment of citizens, cannot be reduced to the problems of democratic participation. That is, simply increasing the participation of citizens in democratic decision-making is no guarantee that they will act responsibly, motivated by a concern for the ecological common good. While there is a greater chance that the quality of social–environmental decisions will be better in terms of ecological rationality under deliberative democratic conditions than under liberal democracy, as Dryzek (1987; 1995) and others (Christoff, 1996; Dobson, 1996a) argue there is no guarantee of this. Just as the problems of democracy cannot be solved simply by more democracy, the question concerning citizenship cannot be dissolved into those of democratic participation itself. As Held notes elsewhere,

> while the evidence certainly indicates that we learn to participate by participating, and that participation does help foster – as Rousseau, Wollstonecraft and J.S. Mill all contended – an active and *knowledgeable* citizenry, the evidence is by no means conclusive that increased participation *per se* will trigger a new renaissance in human development. (1987: 280, emphasis added)

In this sense we may say that citizen participation and deliberation, while necessary, are by no means a panacea for solving social–environmental problems. Given the uncertainty which surrounds the latter, there is nothing which can guarantee their resolution, although the possibility of *transforming* unecological preferences in the light of debate is a necessary condition. Hence the concern with seeing 'green citizenship' as a 'mode' rather than purely a political role composed of a particular complex of rights and duties. As a responsible mode of acting, green citizenship goes beyond the political sphere of relations between state and citizen.

As outlined in the next section, green citizenship is related to what Tocqueville called the 'spirit of association', those associations in civil society within which the virtues necessary for green citizenship can be learnt. It is also the case that responsible citizens need to be socialized not just within the associations of civil society but also within the public education system, as Gutmann (1987) argues from a liberal perspective. The virtues of green citizenship may be learnt and fostered within the public system of education. In the latter case a green state promotes, through public institutions, a green conception of citizenship, just as a liberal state promotes a liberal conception of citizenship. Other state-based fora for socializing green citizens include compulsory public service, as opposed to military service, which could include environmental projects as is the case in some European countries. Such republican-type proposals would seek to create responsible citizens in a way in which public education can only encourage. In other words, there is a degree of compulsion within republican arguments for responsible citizenship which is greater than that found within liberal arguments. This is brought out clearly in Oldfield's frank admission that, from a republican viewpoint, 'The moral character which is appropriate for genuine citizenship does not generate itself; it has to be *authoritatively inculcated*' (1990: 164, emphasis added). Whether green politics goes as far as this is a contentious point, dependent upon empirical conditions. One of the most salient of the latter include the general social perception of the severity of social–environmental relations. It is easy to see how a general perception that these relations have reached a point which threatens social survival may underpin eco-authoritarian arguments for the forcible creation of green citizens. The green democratic defence of responsible citizenship lies, as indicated above, in the normative indeterminacy and epistemological uncertainty which characterizes social–environmental relations, and not just in the necessity for citizens to fulfil duties relating to those relations.

Green citizenship refers to the fact that citizen activism, deliberation, participation, compliance and agreement are required and possible at both the 'input' and 'output' stages of the public policy process. The possibility of increased citizen involvement in making decisions is related to the argument mentioned earlier concerning the diminished but still important role of the state within the context of an ecological redefinition of modernization and progress. Less complex social organization, as a necessary consequence of ecological modernization, implies less need for a centralized state with large bureaucracies and increased opportunities for citizens to take responsibility for their own affairs.[29]

To combat the 'arrogance' of an excessive anthropocentrism, green citizenship seeks to undermine the presumption that an appeal to human interests is sufficient to justify any environmental decision. That such decisions may affect the interests of non-humans must, from a green democratic viewpoint, be taken into account. The point is not that green citizenship demands that humans give up their interests in deference to those of non-humans. Rather, green citizenship is concerned with separating 'serious' from 'trivial' interests, and then specifying the agreed limits within which those interests may be realized. In part, it also requires seeing ourselves as 'citizens-in-society-in-environment', that is recognizing the relations of interdependence, depedence and vulnerability that, as 'natural and social' beings, we find ourselves within. Thus an ethics of use is to be distinguished from 'strong anthropocentrism' which was defined as the claim that human–nature relations can be justified by reference to human preferences alone. This point was made earlier where it was argued that the problem with strong anthropocentrism is that it insulates preferences from critical appraisal. Green citizenship, by contrast, is the praxis of citizens critically evaluating preferences and attempting to come to agreement on limits within which particular social–environmental relations may be pursued.[30] Thus citizenship within the context of green democratic theory and practice is centrally concerned with the elaboration and internalization of a publicly agreed 'ethic of use' for the environment, as opposed to a putative 'environmental ethic'. This idea of green citizenship differs from that of Christoff who argues that, 'To become *ecological* rather than narrowly anthropocentric citizens, existing humans must assume responsibility for future humans and other species, and "represent" their rights and potential choices according to the duties of environmental stewardship' (1996: 159). The emphasis on the rights of non-humans would seem to imply that Christoff's theory assumes an agreed environmental ethic which specifies the rights of non-humans. Green citizenship on his understanding is partly constituted by the discharging of human duties related to those non-human rights. The view of green citizenship developed here, while stressing the importance of duties and obligations, is *not* premised on non-humans having rights. Indeed within this conception of green politics, although it is anthropocentric, there is no *right* to development which may clash with any attributed rights of non-humans. The emphasis on rights talk within green moral theory expresses a proprietarian view of morality which is problematic. However, the appeal to rights in social–environmental moral deliberation may be taken as an indication of the seriousness with which certain views are held, and as such need to be taken seriously. That is, the use of the moral idiom of rights may *indicate* rather than itself *demonstrate or prove* the seriousness of the moral case under consideration. At the same time, it is not to say that rights and duties play no part within the green conception of citizenship. Collective ecological management does, in building upon key aspects of ecological modernization, require a transformed ensemble of legal rights and duties between citizen and state (Weale, 1992: 150). Viewed in this way the emphasis on deliberative citizenship here can be seen as a necessary and desirable complement to the 'green state'.

The deliberative character of green citizenship can be understood as relating to the variety of forms of human valuing in relation to the environment, and the multiplicity of values that mark human–nature relations. The deliberative, communicative understanding of democracy and citizenship within green politics can be taken as a necessary consequence of its refusal to accept that one form of human valuing (such as an economic one) can regulate any social–environmental metabolism. It was partly for this reason that economistic views of social–environmental relations were criticized since they narrow the range of human interests in the world.

This is not to say that economistic reasoning and valuing are to be excluded from the determination of social–environmental interaction. In the case of non-humans, the deliberative democratic process allows the *representation* of their interests as perceived by citizens or political institutions. Economistic forms of valuing typically narrow human interests in social–environmental affairs while also denying the interests of non-humans to have any bearing on decision-making. Economistic forms of valuing in practice *demoralize* social–environmental interaction, seeing the latter purely in terms of a material-cum-economic transaction with the satisfaction of a narrow set of human interests as the only justification needed. Green citizenship is to be understood as a corrective practice to the 'vices' of an arrogant anthropocentrism. At the same time green democratic citizenship may also be considered as part of the process through which an expansion and redefinition of the 'economic' can be effected, as indicated in the last chapter. Green citizenship *qua* ecological stewardship cannot be confined to the 'political sphere', narrowly understood as referring to the state. It denotes a particular constellation of rights and duties which range over spheres of social and private life which impact on the environment. Just as the conception of the 'economic', which is the main subject of analysis for green political economy, is much broader than that given by 'formal' economic transactions, so green citizenship, while based within the formal political sphere, is not confined to that sphere. Being a good citizen from a green point of view consists not merely in considering the interests of non-citizens in making environmental choices, but also in acting in a manner which promotes ecological stewardship. The point about ecological stewardship is that it is a particular human social *mode* of being, acting, thinking and feeling.

As a practice, green citizenship is the ethical core of collective ecological management, the broad institutional framework that regulates social–environmental relations. And although the primary locus of green citizenship may be territorially defined within the nation-state, the latter does not delimit its scope. Given the transnational character of environmental problems, green citizenship is guided by the green slogan of 'act locally, think globally'. The ethical and prudential dimensions of stewardship are not and often cannot be expressed at the local level but demand an integrated approach which combines local, regional and global dimensions. The challenge that green political theory proposes for the contemporary arrangement of the global human community is to institutionalize politically the moral concerns expressed by ecological stewardship. This is related to the argument in Chapter 3 where the idea of 'citizen-in-society-in-environment' was used as a way of expressing the shift in perspective registered by green politics. Since the environment not only includes but also transcends the nation-state, green citizenship opens out new arenas for citizen activism. As Christoff notes:

> The citizen's political community (which, for many other issues may remain that of the nation-state) is profoundly reshaped by an ecological emphasis which generates additional and occasionally alternative transnational allegiances ranging from the bioregional to the global, as well as to other species and the survival of ecosystems. (1996: 159)

The transnational character of green citizenship can be taken as a political expression of the increased ecologically based interdependence that creates new relationships between otherwise unconnected individuals. While green citizenship is state-based, its role as a mode of ecological stewardship means that it cannot be confined to the state. To be a good green citizen requires expanding one's sphere of action to transnational and even global levels.

Civil Society and Green Democratic Theory

In this section I discuss the place of civil society within green democratic theory. The incorporation of the concept of civil society into green political theory further distinguishes the conception of green theory being developed here from the eco-anarchist position, discussed earlier. Firstly, the adoption of the state/civil-society perspective explicitly rejects the retreat into *Gemeinschaft* that typifies bioregional politics. To adopt a state/civil-society perspective is to have an understanding of society as *Gesellschaft*, society as 'corporate association' rather than as 'community' in the bioregional sense. Secondly, the concept of civil society stresses the organization and regulation of the economy as central to determining the character of the connection between state and civil society. This was a central aspect of both ecological modernization and collective ecological management. Regardless of which interpretation of civil society one takes, both 'liberal' and 'post-liberal' conceptions (outlined below) regard the organization of the economy as of central political significance. The centrality of the economy to the green analysis is obvious, as it is within the economy that the material exchange between society and its environment occurs, and within which one can find the origins of most environmental problems. The material metabolism between economy and environment can thus be regarded as the primary site of the cultural contradiction that is often misleadingly termed the 'ecological crisis'.

The concept of 'civil society', like many other popularly used terms within political theory discourse, has a high level of usage but a marked level of disagreement as to its precise understanding. For present purposes, following O'Neill (1993: 177), we can outline two senses of civil society. On the one hand there is what one may call the 'traditional liberal' understanding of civil society which identifies it in the first place with the market society that first emerged in the eighteenth century and developed throughout the industrial period of Western societies (Polanyi, 1957). In this understanding of civil society, the market order is regarded as a key basis upon which the freedoms of civil society can be secured. An economy regulated by the market is seen not simply as the most efficient organization of the economy, but also as a necessary bulwark against excessive state interference and totalitarianism. On this view, a planned economy expresses a collectivist totalitarianism, the negation of the liberal order and individual liberty (Hayek, 1976).

The other conception of civil society, which may be called a 'post-liberal' view, regards it as referring to associations that are independent of the state and the market economy (Keane, 1988). On this view, the freedom of civil society is to be found in the autonomous practices of individuals and groups within civil associations, non-market institutions such as professional bodies, voluntary associations, clubs and societies, as well as institutions that are funded by the state but are not of the state such as universities, schools and hospitals (O'Neill, 1993: 179). On this view totalitarianism does not come about as the result of a planned economy; rather it is the state's abolition of civil associations and the 'public sphere' which secures totalitarianism. There is a presumption that such a conception of civil society needs to be protected not just from the state, but also from the corrosive effects of the market economy (1993: 181). It is this second understanding of civil society that is closest to green concerns. This can be seen in the emphasis laid on the 'public sphere' within many discussions of green politics (Dryzek, 1995; Christoff, 1996; Dobson, 1996a; Eckersley, 1996b), the green movement's association with the 'new politics' of civil society (Melucci, 1989; Doherty, 1992), the critique of the market developed in the last chapter, and the critique of overly statist, administrative approaches in Chapter 5.

Green democratic theory is concerned with the relationship between a democratic society and its political system, a central aspect of which turns on the separation of the state and civil society. This is expressed by Keane as implying that 'the separation of the state and civil society must be a permanent feature of a fully democratic social and political order' (1988: 13). Without this division democracy is impossible, and without democracy this division becomes difficult, if not impossible, to sustain. The institutionalized separation of the state and civil society is therefore a constitutive part of any defensible democratic political order, green or otherwise. Since the separation of state and civil society is, as

Held notes, a 'fundamental liberal notion' (1987: 281), the acceptance of this division by greens is further evidence of the 'post-liberal' complexion of green democratic theory.

Part of the new pattern of relationships between state and civil society that concur with green thinking includes suggestions for the state to '"lease back" institutions of social policy to the community' (Held, 1987: 288; Keane, 1988). The institutions of collective ecological management are obviously open to such an option. Other welfare institutions which if leased back would concur with the central aims of green democratic theory include institutions of medicine, housing and education. Before the new right and non-statist socialists adopted it, green theorists such as Illich (1971; 1973; 1974; 1975), Goldsmith (1972a), Robertson (1983) and Gorz (1983) had articulated a critique of the contemporary welfare state on the grounds that it undermined individual and collective autonomy and self-reliance. Illich's arguments concerning the 'radical monopolization' of core areas of personal and social life by professional agencies was at root a critique of the expropriation of the definition of needs as part and parcel of the process of economic modernization. Together with Gorz, Illich has been at the forefront of developing green arguments for the autonomy of civil society from both the market and the state and the restructuring of relations between them. A central part of this restructuring process from a green perspective involves shifting decision-making power in regard to 'social welfare' away from the market and the state and returning it to individuals and communities. This fundamental restructuring of the definition (ends) as well as the institutions (means) of welfare is at the heart of the green aim to place the market and the state at the service of civil society rather than vice versa (Barry, 1998a). Thus the position outlined in the last chapter, which stressed the significance of economic self-reliance, the local economy and definitions of welfare, can be regarded as the political economy underlying the restructuring of relations between state, market and civil society. This aspect of green theory could be viewed as shifting from a 'politics of social welfare' in which the market and the state define and administer to the imputed needs of civil society, to a 'politics of well-being' in which individuals within civil society define their own needs to a greater extent. Social well-being, from a green perspective, is concerned with individuals not as consumers or clients but as social beings, whether this sociality is expressed through membership of, and participation in, communities or civil associations. What is meant by this is that a green conception of individual well-being sees it as set not simply within a social context, but within social *practices*. In other words, human well-being from a green perspective (the 'green view of the good') consists in *doing* rather than consuming or having. And while social practices do of course have a material impact on the environment, it is clear that a view of human well-being in which the emphasis is on social (which can include political) interaction as a major, though not the sole, component of the good life will make fewer demands on the environment.

Work as Social Practice and Economic Activity

An example of the practice-based view of the good life is the green argument for work to be transformed into an intrinsically valuable social activity, done, as far as possible, for its intrinsic benefits rather than individual monetary remuneration and capital accumulation. This implies a restructuring of the economy which would transform the regulative goal of the economy away from narrow economic efficiency and accumulation, and seek to integrate these 'economic' goals within wider social, moral and ecological considerations. Part of this, as suggested in the previous chapter, would involve restructuring the formal economy to permit the internal goods of work to be realized, as well as expanding the idea of the 'economy' beyond the formal or monetized economy. According to Gorz (1989), the destruction of work as a social practice is tied up with the rationalization of labour under capitalism. For him,

> The economic rationalization of labour . . . was a revolution, a subversion of a way of life, the values, the social relations and relations to Nature, the *invention* in the full sense of the word of something which had never existed before. Productive activity was cut off from its meanings, its motivations and its objects became simply a *means* of earning a wage. It ceased to be a part of life and became the *means* of 'earning a living'. (1989: 21–2)

In this way, reversing this process and viewing work as a social practice, rather than simply as the use of the factor of production 'labour' in the formal economy for the production of commodities, would contribute towards the creation of a more sustainable economy–ecology metabolism. This transformation of work is similar to Lee's arguments for a less productivist and more aesthetic or craft mode of labour (1989: Chapter 8) within what she calls a necessary and desirable 'deindustrializing' process (1993a). It also draws upon Robertson's (1983) 'post-industrial' prediction of a future in which the proportion of economic activity made up of 'ownwork' will increase. This is because the prevailing definition of work as paid employment in the formal economy has now come to an end, and 'full employment' is a thing of the past. Both Lee and Robertson provide arguments for the position outlined in the last chapter for the green ideal of self-reliant 'prosuming': defining, producing and consuming what one needs outside the market and the state as much as possible, while not rejecting the opportunities for trade that the market affords, or the meeting of needs that the state can provide.

This redefinition of work is thus one important aspect of the green aim to 're-embed' the economy within society as a necessary step in the creation of an ecologically rational sustainable economy–ecology metabolism. Recalling that ecological rationality has normative as well as material dimensions, this reconceptualization of work concurs with the general green aim to 're-moralize' human/non-human interaction within the context of an ethics of use. This can be most clearly seen in relation to work which involves human interaction with animals. Here the green argument for the creation of more 'human-scale' and personal forms of economic practice, which is an underlying aim of the green restructuring of work, converges with the normative aim of remoralizing human/non-human relations. Supplementing and supporting the critique of intensive livestock rearing in Chapter 3 on the grounds that it privileges economic over other human interests in the environment, we now have the argument that less 'industrial' forms of the human use of animals can permit moral considerations to regulate that use.

While of course there is no guarantee that a less industrial and large-scale human use of animals will result in a more 'humane' metabolism, or that this reconfiguration of human–nature practices can be extended to other human uses of the non-human world, the green argument is that such a reconfiguration is a necessary feature of the remoralization of human/non-human interaction. The green contention is that returning work to the category of a 'practice' which stresses the internal goods associated with it may, in the case of human use of animals, permit the possible realization of 'ecological virtues' associated with stewardship. In other words, viewing work as a practice would constitute the necessary social re-embedding of economic activity required for the latter's integration within its ecological context, as concluded in the last chapter. The more social–environmental purposive, transformative relations approach the ideal of a social practice, the more that use realizes the virtues of stewardship rather than exploitation.

At the same time, this view of work recasts it as a central site for the cultivation of ecological virtue. That is, work is no longer an 'impersonal' activity in two senses. Firstly, this reconceptualization of work is one form that the 'repersonalization' of human–animal productive relations that Benton (1993) seeks can take. Secondly, work is conceived as a site of character formation, within which habits and virtues can be cultivated, and not just an activity engaged in for monetary remuneration. Thirdly, in keeping with the argument in the last chapter, work is to be regarded as a way in which citizens are socialized, or given a stake in society, and included as valued members of, and contributors to, society. As Jacobs points out,

> We should deal with both sides of the coin – not just with the absence of work, but with work itself . . . Society's goal should therefore not be simply to increase employment. It should be to enable people – both men and

> women – to do the variety of kinds of work which provide benefit to themselves and to society . . . The longer-term goal should be to reduce 'normal' full-time working hours. (1996: 87–9)

In redefining the institutional boundaries of the relationships between market, state and civil society, the green argument for the reconceptualization of work can be regarded as an attempt to protect, as far as possible, the *practice* of work, its internal goods and virtues, from *institutional* distortion in the form of market- or state-imposed external goals, while recognizing the continuing importance of formal employment. As mentioned above, a principal external goal that greens are keen to restrict and reformulate is the imposition of policies aimed at economic modernization, whether they be state- or market-based. In short, under economic modernization 'work' as a *social* activity loses its internal goods, as a result of changes in scale, technology, deskilling, the division of labour etc., and increasingly becomes an *economic* activity.

This distinction between work as a social practice and as an economic market-regulated activity, and the tension between them, are related to the issue raised earlier concerning whether a goal of green politics is social transparency. While it is obvious that some versions of green political theory seek the transparency of social relations, particularly eco-anarchism, the conception of green politics defended here does not. Rather, it accepts that there will be social–environmental productive relations within which the ideals of work as a social practice will be compromised by work as an economic–institutional activity. One way of looking at this is to see that those productive-transformative relations mark the sphere of social–environmental exchanges within which external goods and criteria, such as productivity and efficiency, may be legitimately pursued without violating the ethics of use. Human productive use of the environment beyond basic need fulfilment is legitimate. However, such productive relations may mark the limits of acceptable instrumentality: for example, luxuries seem to occupy a permanent position on the border between 'use' and 'abuse'.

This redefining of work as a social practice embodying the purposive human transformation of the non-human world can also be seen as compatible with the argument outlined in the last chapter concerning how the gap between production and consumption can be lessened by encouraging self-provisioning and ownwork. The underlying normative justification of the reconceptualization of 'work' is that such a reconceptualization is the keystone in the reorientation of social–environmental relations. Work, perhaps along with transport, eating and food production more generally, is the activity which expresses the central features of the social–environmental metabolism. Changing the normative parameters within which this core social–environmental practice takes place is thus a, if not *the*, central political-normative goal of green politics, one that would have cultural as well as political and economic repercussions. That is the 'greening of work' is a keystone issue for green politics in that it is not only central to the creation of an ecologically rational economy–environment metabolism (the 'greening of the economy'). Given the significance of work culturally and politically it is also, and will continue to be, a central aspect in the 'greening of society'. Changing the character of work is thus a central task in changing the character of modern society and the modes of experience and character of its citizens.

Conclusion

In a prescient quote, Lynn White suggested in 1967 that, 'Our ecologic crisis is the product of an emerging, entirely novel, democratic culture. The issue is whether a democratized world can survive its own implications' (1967: 1204). The conclusion of this chapter is that a 'democratized world *can* survive its own implications'. Against the eco-authoritarian position, green politics is not necessarily, however regrettably, convinced of the necessity of an anti-democratic stance. At the same time, this chapter has also sought to establish that the complexity and difficulties associated with social–environmental problems do suggest that those greens who place their faith in direct democracy are equally mistaken. It is *not* the case that the ecological problems of democracy can be solved simply by *more* democracy. *Better* democracy may be a necessary condition for enhanced ecological rationality, but even this is insufficient to guarantee a sustainable economy–ecology metabolism, never mind a symbiotic social–environmental one. Better democracy is to be understood as having to do with the importance of democratically constructing a more ecologically rational culture which is partly constituted by green citizenship and 'work' as a social practice. Green citizenship views citizens as sharing responsibility for environmental protection, together with the state. At the same time, these duties are balanced by citizen rights with regard to environmental decision-making and the democratic accountability of the policy-making process. Only such a cultural transformation can provide anything approaching a 'guarantee' of sustainable and symbiotic social–environmental relations, which while aiding the transition from ecologically irrational modes of interaction, attempts to sustain the positive gains associated with those modes. The aim is not to save the world above all else, or secure the infinite continuation of the human species, but rather to stop and reflect upon what it is we are doing to both ourselves and the world, and to search for a new way forward. The point is that our dependence upon the earth (and to an growing extent 'its' dependence upon us) is both increasingly obvious and makes a new, updated version of stewardship the most defensible form that dependence can take. The challenge is to achieve this while maintaining and adapting democracy as the most defensible form our collective political dependence on one another can take.

Notes

1 Classical liberals such as Tocqueville assumed a relationship between an affluent economy and political democracy. One aspect of Tocqueville's thought turns on the idea that 'a flourishing economy is essential to the stability of democracy, since it gives defeated politicians an alternative, which makes them more likely to accept defeat rather than attempt to illegally hold on to office' (Copp et al., 1995: 3). Classical Marxism, on the other hand, assumed a connection between 'emancipation' and material abundance. The roots of the different understandings of the connection between the two may lie in the interrelationship between the logics and legacies of the industrial and French revolutions, understood as expressing the core values of modernity, one relating to economic abundance and the other to political democracy.

2 While the 'terrible trio' are routinely wheeled out for ritual condemnation within the literature (Eckersley, 1992a: 11–17), other authoritarian aspects of green thinking, such as that which can be detected within aspects of deep ecology, go unnoticed. See Barry (1994: 377–8) and Vincent (1993: 266).

3 There is a similar argument regarding the relationship between population growth and economic development. The 'demographic transition', which is usually understood as the decrease in population growth as a result of economic growth, is more a function of economic security than of growth and affluence. According to Sen (1981), it was not the wealth of European capitalist societies which caused a drop in population and staved off famine, but their welfare systems, which guaranteed all citizens a minimum bundle of welfare goods and services. Similarly, one can argue that it is the *distribution* of wealth within society, not the *absolute level* of wealth, which is important in a democratic political system. Hence green concerns with minimum levels of socio-economic welfare (as in the basic income scheme), as well as a concern with placing limits on maximum income levels (Daly, 1973b).

4 Realistically, it may mean that a more redistributive political order will be required. As discussed later, a more egalitarian redistributive order would undermine justifications for socio-economic inequality premised on the latter's necessity (in terms of efficiency) in procuring economic growth. It is, in large part, an empirical question as to whether a more egalitarian distribution of wealth, opportunities and resources would be a more ecologically sustainable one, but there are good reasons to expect that it would.

5 Thus the oft-cited anti-scientism of green political theory may have more to do with the undemocratic implications of decision-making by experts than anything else. See O'Neill (1993: Chapter 9) on the democratic preconditions for submitting to the authority of science.
6 In this respect Ophuls's position is in many ways a modern articulation of the argument for a 'benign dictator', in that Ophuls argues for macro-level coercion and centralized authoritarianism but with micro-level freedoms and democracy (Mills, 1996: 98).
7 Though there are some writers on green politics who wish to translate spirituality into politics, such as Skolimowski (1993) and Mills (1996: 113).
8 For example, public suspicion and concern about genetic engineering seem to be centred on the secrecy surrounding the development of this technology as much as on the ethical misgivings people may have of humans 'playing God' in designing new life forms.
9 Evidence for this may be observed in the explicit interdisciplinary or transdisciplinary approach that is required in thinking about dealing with social–environmental issues. Social–environmental relations span the natural and social sciences and humanities, and no one discipline (let alone a sub-discipline) can pretend to offer a comprehensive epistemic basis for thinking about social–environmental relations.
10 An almost identical position has been articulated by Kai Lee who holds that, 'Because human understanding of nature is imperfect, human actions with nature should be experimental. Adaptive management applies the concept of experimentation to the design and implementation of natural-resource and environmental policies' (1993: 53).
11 This acceptance of uncertainty can also be seen as underwriting prudence as an ecological virtue. The persistence of uncertainty as a background condition, particularly within expert systems of knowledge, may be taken as a breakdown in the 'arrogance' of humanism discussed earlier, and may indicate the emergence of a more humble (less 'arrogant') conception of the relationship between humans and nature. Accepting uncertainty as a permanent feature of social–environmental relations is one way of recognizing the limits to human knowledge of the world. That we cannot know everything about the world may turn out to be a necessary rather than a contingent condition, given that the world is always revealed 'for us' rather than 'in itself' as Kant noted. And this applies *a fortiori* if we follow O'Neill's (1993) wise advice and recognize that pursuing the 'truth' is not the only or even the highest human good.
12 For the only book-length exploration and defence of liberal democracy on ecological grounds, see Wissenburg (1998).
13 In the eighteenth century market-led economic modernization was defended and promoted partly on the grounds of its 'civilizing' effects. The green critique may be viewed as holding that economic modernization, particularly in its present globalized form, may actually have a 'barbarizing' effect on social relations, and an unsustainable and unsymbiotic effect on social–environmental relations.
14 An alternative argument has been proposed by Lee (1993a), who argues that it is rational to argue for deindustrialization on the grounds that it is not only ecologically beneficial in terms of sustainability but also positive in terms of

restoring the moral worth of the non-human world and the dignity of labour by permitting a more 'artistic' mode of production.

15 One way of looking at this is to say that collective ecological management, while being state-based, is not state-centred, as it requires non-state institutions such as the (local) formal and informal market and the institutions and practices of civil society.

16 According to Lauber (1978) there is evidence to show that the relatively liberal, and consequently less powerful, British state was an important determinant of the stagnation and decline of its economy since the Second World War. Relying on the comparative studies of Schonfield (1965), he states that, 'the governments that have been most successful in the pursuit of the new [economic] goals have been those which had few doubts about the extensive use of non-elected authority, for example, France. The more "timid" governments were less successful' (1978: 209). Having 'modernization' as one's highest goal may lead to non-democratic, illiberal forms of state action.

17 More recent examples of non-democratic paths to economic modernization can be seen in the South East Asian 'tiger' economies which combine capitalist economic organization with authoritarian and non-democratic 'traditional' political structures.

18 Although beyond this chapter, there is reason to think that as well as different democratic institutions, there may be different democratic decision criteria that a comprehensive account of the democratic regulation of social–environmental relations would have to address. For example, it might be that simple majorities will not be acceptable for deciding a referendum on a national 'sustainability plan', where a two-thirds or three-quarters majority (or perhaps consensus) is more appropriate.

19 This distinction between market, state and community as three possible mechanisms for the regulation of collective life is taken from Taylor (1982: 59).

20 It must be said however that other conceptions of green political theory which stress the superiority of the 'natural' over the 'unnatural' or 'artefactual' do seem to orientate themselves towards discovering some objective truth of human–nature relations, the 'proof' of which is to be found in the infinitely sustainable character of the metabolic relation which follows from that 'truth'. This was discussed in respect to deep ecology in Chapter 2, and its 'submissive' or reverential attitude to nature. For an example of a conception of green politics in which the value of the 'natural' is central see Goodin (1992).

21 The underlying rationale is that the decision to have children cannot be taken to be a purely private matter but, because of the social–environmental impact of population growth, must be seen as a private act which will have public effects. As such the decision to have children can be viewed as a potential externality which raises issues concerning the balance between the protection of the individual from social coercion and the protection of the public good of a sustainable social–environmental metabolism. Although there are many precedents for the democratic regulation of ostensibly private behaviour, such as laws concerning driving and the production and consumption of certain goods and services, the regulation of population increases does pose some especially difficult problems. Being denied the

right to drive as fast as one can is not the same as being denied the right to have as many children as one wishes. While the political regulation of private decisions concerning procreation will form a part of any green democracy, this need not be as large a problem as it seems if one views production and consumption decisions as having a greater or equal weight in affecting the environmental impact of social relations. On this account, it is ecologically rational, not to say more politically acceptable, to regulate production and consumption rather than population. From a global perspective, focusing exclusively on population tends to omit the issue of per capita consumption or per capita environmental impact. However, as suggested in the last chapter, the aim of ecological stewardship as an ideal or virtue is to integrate one's roles as parent, producer, consumer and citizen into a mode of character, of acting in the world, in which decisions to have children will be based on interests other than one's own. This is discussed in the next section.

22 The incorporation of the interests of the future would be akin to interests of children being entrusted to parents. Another model is that of many aboriginal peoples for whom each major social decision is assessed in the light of how it will affect future generations. Such sentiments are expressed in the native American saying, 'We do not inherit the earth from our parents, but borrow it from our children.'

23 It needs to be stressed that the interests in question here are welfare not liberty ones. Recalling the distinction made in the last chapter between 'welfare' and 'liberty' considerations, democracy from a green perspective is justified on the grounds that it can secure basic liberty interests, particularly that of autonomy. In other words, green democracy is defended on deontological rather than utilitarian grounds. As such, democracy within green politics is conceived as an intrinsic rather than an instrumental practice. That is, it is valued as a procedure for making political decisions, including social–environmental ones.

24 Part of the reasoning behind this is that behavioural changes motivated by the internalization of norms are more effective and longer lasting than behavioural changes based on external or coercive imposition. This suggests a critique of the eco-authoritarian position on the grounds of effectiveness, premised on the assumption that change motivated by an acceptance of its moral rightness is more effective in sustaining that change than if that change is grounded in fear or coercion. The state cannot do everything. As Cairns and Williams suggest in another context, 'What the state needs from the citizenry cannot be secured by coercion, but only by co-operation and restraint in the exercise of private power' (1985: 43) – responsible citizenship, in other words.

25 One example frequently given to demonstrate that the process of aggregating preferences cannot be extended to political decision-making is the widespread phenomenon of individuals supporting institutional forms from which they may receive no personal benefit. For example, many people support non-entertainment public broadcasting or the public support of so-called 'high culture', even though they do not watch those programmes or enjoy 'high culture'; or they support strict environmental standards or the protection of species, despite the fact that they may not derive any benefit

from such legislation. Such examples are evidence of 'public good thinking' discussed earlier.

26 It also echoes Barber's view that democratic citizenship is 'the only legitimate form our natural dependency can take' (1984: 104). On this view the green argument is that democracy is the only defensible form our (human) dependence upon the non-human world can take.

27 This account of green citizenship differs somewhat from that proposed by Christoff (1996). For him it is an 'emancipatory project which is shaped by – and in turn constitutes – ecological citizens' (1996: 162). Now while not ruling out this possibility, green citizenship as developed here has to do with the human-centred concerns of ecological stewardship, not the emancipation of nature.

28 This virtue of green citizenship is close to the liberal virtue of what Galston calls 'the virtue of public discourse', which he defines as including 'the willingness to listen seriously to a range of views . . . The virtue of public discourse also includes the willingness to set forth one's own views intelligibly and candidly as a basis for a politics of persuasion rather than manipulation or coercion' (1991: 227). As it stands Galston's defence of liberal political theory, which in many ways harks back to the 'social liberal' tradition and a 'developmental' view of democracy, stands at odds with contemporary liberal democratic practice which is mainly concerned with the aggregation of individual preferences.

29 A general reconfiguration of modernization (Beck, 1992) or more radically a wholesale process of deindustrialization (Lee, 1993a), which would undermine arguments for the necessity for a centralized, highly bureaucratic state, does not, as Carter (1993) argues, lead to the abolition of the nation-state and the establishment of an eco-anarchist political structure. An alternative, and one that will be canvassed in the section on civil society below, is that such an ecological reorientation of contemporary society heralds a new relationship between state and civil society.

30 Although green citizenship as a form of ecological stewardship appears close to deep ecological notions of 'ecological selfhood', particularly with regard to the internalization of the interests of non-humans, there are significant and important differences. Ecological stewardship lies in the public, collective determination of the rights, duties and personal qualities of citizens with respect to the achievement of an ecologically rational social–environmental metabolism. Ecological selfhood, as discussed in Chapter 2, lies in the largely private sphere of intuition and revelation. It has little or no 'political' and by implication 'democratic' dimension, either in its determination or in its expression, unlike ecological stewardship. The standards or virtues of stewardship are intersubjectively created, not objectively given.

8

Conclusion: Nature, Virtue and Progress

CONTENTS

The aim in this book has been to rethink green politics, to outline the contours of a green political theory which, while being consistent with the values and principles of green ideology, explores them in a broader, less constraining context. In keeping with a critical-reconstructive approach, I hope to have suggested a conception of green political theory which greens can identify with, if not fully endorse, while at the same time presenting green arguments which can appeal to non-greens concerned about social–environmental problems. While by no means presenting a fully fledged, complete account of green political theory, this book is intended as a contribution to the 'theoretical consolidation' of green politics mentioned in the introduction.

One thing which this analysis demonstrates is how much more difficult it is when the centre of gravity of green politics moves from a rejection of the status quo guided by reference to some future social order, to a critique based on an analysis of the principles underlying the status quo and working through their implications from a green perspective. Part of this shift within green theory requires an engagement with standard topics and themes within political theory, such as democracy, equality, autonomy and the state. It also has to do with the character of green political theory being tied up with distinctively 'green' political issues such as ecological sustainability. That is, green political theory suggests legitimate new concerns for political theory, as well as reviving older ones and giving them a new relevance.

In the rest of this conclusion I want to focus on two aspects of rethinking green politics which deserve to be highlighted. These are the idea of 'progress', and virtue *qua* ecological stewardship.

Green Politics, Nature and Progress

One of the defining themes of green political theory, and one which is so obvious that it often goes unremarked, is its attitude towards and concern with 'progress'. One way to view green politics is to see it as a critical reaction to modernity, and its particular idea of progress. More specifically, one can view green politics in terms of its attitude to the two revolutions at the heart of modernity, namely the industrial and French revolutions. These two revolutions are shorthand ways to refer to two of the central 'logics' of modernity: socio-economic advancement or development, and democracy, equality and solidarity. It is the dialectic between these two logics that forms the historical origins and the theoretical dynamic of green political theory. In standard ideological accounts, green politics criticizes and/or rejects the industrial revolution and seeks to extend the 'democratic project' initiated by the French revolution.[1] However, as a political theory concerned with the relationship between the industrial and French revolutions, the question of progress is one that green political theory cannot avoid addressing. This is not simply in terms of substantiating its 'progressive' self-understanding. It is a key issue within the process of rounding and fleshing out the character of green political theory itself.

The whole tenor of early and ideological accounts of green politics resonates with a perception that the costs of modernity outweigh the benefits, which are themselves suggested to be of questionable quality. From this position two equally unappealing understandings of green politics can be advanced. On the one hand, green politics constitutes a rejection of modernity's legacy of progress in both the socio-economic and political spheres, that is, green politics is both anti-industrial and anti-democratic (which is close to the 'romantic-conservative' reaction to modernity). On the other, green politics implies a rejection of industrial progress but the acceptance and indeed radicalization of 'democratic progress'. This more popular view of green politics sees it as anti-industrial, but pro-democracy. Now while the latter has obviously more to commend it than the former, I have suggested that the anti-industrial tenor of green politics, within which is subsumed the common rejection of consumerism, materialism, science, technology and the market economy that marks much green writing, needs to be questioned. Finally, rethinking green politics as an 'immanent critique' of modernity avoids it being viewed as either 'pre-modern' or 'post-modern'.

Gowdy (1994) represents the type of green view of progress which I wish to criticize.[2] According to him

> there is no convincing evidence that past economic growth has led to unambiguous improvement in the human condition. Once we give up the idea of progress, we can concentrate on the making do with what we have rather than placing our hopes on some future material or ethical utopia. (1994: 55)

His injunction to abandon the idea of progress would, I argue, be a retrograde step, not just conceptually, but also practically. It is only if one equates progress with *undifferentiated* material economic growth that it makes sense from a green position to talk of abandoning progress. But progress does not necessarily equate with economic growth, as writers from J.S. Mill to more sensible contemporary green critics such as Beck (1995) have emphasized. Rethinking green politics requires redefining, recalibrating and reappropriating, rather than rejecting, the politically powerful idea of progress. Gowdy's extremely pessimistic advice to greens is that when they are accused of being 'anti-progress', to turn the tables on their opponents by rejecting the assumption 'that progress has taken place' (1994: 55). To claim that no progress has taken place since the Enlightenment is not only foolish but dangerous advice. Progress has taken place, albeit unevenly, unreflexively and, up until now, largely without concerns for global or local sustainability or symbiotic moral relations with the non-human world. This rejection of progress is neither necessary nor desirable for green arguments, and highlights the stark gulf between those who see the ecological crisis as a 'total crisis' and those who see it as a contradiction within contemporary advanced societies. After all, problems with a particular conception and institutionalization of progress do not demonstrate the worthlessness or wrongheadedness of pursuing progress. In this respect, the form of green politics outlined here may have some connection with Habermas's idea of the 'unfinished project of modernity'. Green politics, in offering an immanent critique rather than a rejection of modernity, may be an essential part of delivering its promises.

Against this view, I have sought to show that green political theory is premised on the redefinition of progress, understood as an immanent critique of modernity, not a politics seeking to return to either a romanticized pre-modern social order or a post-modern rejection of the present. While rightly highlighting the costs associated with the industrial revolution and its legacy, to reject all its fruits as 'false' would be churlish, not to say foolish. Thus while accepting the democratic claims of green theory, I have also sought to deepen the immanence (and thereby the practical relevance) of the green critique by seeking to present it as a critique of modernity's legacy of human progress in *both* the political and the social spheres. Thus, I have argued that consumption, materialism, science and a market economy (as opposed to the present global capitalist one) can, and ought, to have a place within green political theory. Particularly in respect to science, green political theory cannot consistently reject modernity since modern sciences from ecology and conservation biology to thermodynamics have played, and continue to play, a central role in its evolution and development. Indeed, part of the novelty of the green political perspective lies in it being the first political outlook so informed by and grounded in modern science. While one is keen to stress the scientific credentials of green politics and to accept that green politics can be seen as part of the 'progressive' narrative of humanity, in the sense of the improvement of our lot as a result of our increasing knowledge of the non-human world, this is balanced by an acceptance of the moral and epistemic limits of this process. Green political theory explicitly recognizes the limits of human knowledge, and accepts that in principle we can never know everything about the world. In that sense, the world will always remain opaque to us; its workings and our relationship to it will never be fully transparent.[3] Alongside this epistemic limit to knowledge of the world is the moral argument that knowledge and the pursuit of truth are not the only or often the most pressing or important social or moral goals. The pursuit of truth is one amongst other values, and it does not enjoy a principled *a priori* pre-eminence, as O'Neill (1993) rightly points out. Therefore the pursuit of knowledge requires moral constraints to be placed on it. And since knowledge is power, with power comes moral responsibility. The tremendous power humans have to affect the natural world demands that moral concern and deliberation characterize its use, lest this knowledge become another ecological vice if divorced from such moral considerations. A final qualifier to the scientific character of green politics is the acceptance of certain 'givens' of the human condition, which implies that scientific knowledge and its fruits are to help cope with rather than eliminate these aspects of the human condition. These 'givens', as discussed in Chapter 3, include the inevitability of death, the central foundational importance of birth, reproduction and collective subsistence, human vulnerability and dependence (all of which we share with other animals), human plurality, human culture, and above all else the dynamic and uncertain character of our dealings with the natural world. The explicit recognition of our epistemic limits and the awareness of our ignorance are of course ecological virtues to help us avoid the ecological vice of an arrogant anthropocentrism, and an exaggerated sense that technology can 'solve' every social–environmental (and social) problem.

While there may be genuine debates about the necessity or utility of enclosing the commons today, green political theory does not need to base this on an *a priori* rejection of enclosure whenever and wherever it occurs. The green point is that while progress was premised historically on enclosing the commons, arguments that future social progress necessitates either further enclosure (which is the neoliberal position) or the continuation of the current pattern of social interaction with the environmental commons are debatable, to say the least. As I hope to have demonstrated, an important aspect of the theoretical consolidation of green political theory requires focusing on matters of institutional design. A central part of this requires seeing commons-type regimes as one institutional choice alongside other distinctively 'modern' institutional forms, namely those of the (formal) market and the nation-state (including transnational and subnational forms of political authority). There are alternatives, as I have suggested, to the idea that returning people to the land (and the land to the people) is the only or most appropriate manner in which to create and maintain an ecologically rational social–environmental metabolism. While one could justify (*ex post*) the enclosures as a necessary precondition for the industrial revolution, green political theory can be read as suggesting that further progress cannot be based on the patterns of past social development.

It is within this context that the green political economy argument in favour of a 'post-development' perspective of Chapter 6 ought to be read. Green political theory can thus be understood as based on a critique of the linear, one-dimensional versions of social progress which equate it with 'economic modernization' after the model of Western industrial societies. Progress can no longer be simply (and simplistically) equated with ever increasing material affluence, the multiplication of desires and consumption- or market-based economic organization. This is the point about the principles of green political economy and how it differs from ecological modernization. Unlike the latter it proposes a different type of progress, a view of development which emphasizes qualitative as well as quantitative indicators or criteria for judging social progress. And most importantly, it explicitly recognizes that the 'ends' of social progress (social progress itself), and not simply the means to it, can and ought to be subject to democratic deliberation. Ecological modernization is a positive step in the direction of this recalibration of progress, but as argued in Chapters 5 and 6 is still within the larger framework of 'economic modernization'. That is, its main aim is to ensure continued economic growth. Perhaps most importantly, ecological modernization's emphasis on state regulation of the market can be read as indicative of the type of institutional innovation required if we are to map out the parameters of a new course for social progress. In this way one can view its contribution to the development of collective ecological management and ecological stewardship in terms of suggesting a shift from ecological *modernization* to something approaching ecological *enlightenment*.

What I mean by this distinction is that progress under ecological modernization follows, in essence, the past patterns of economic development, particularly the equation of economic growth with human social progress. Now the point about the latter 'orthodox' model of progress is that while state-directed economic modernization has always played a greater or lesser part, by and large the assumption of this model is that progress is the unintended outcome of social interaction between individuals and groups with different purposes. This has been the model of social progress which has held sway from the eighteenth century to the present day. As Adam Ferguson (1966: 122) noted in the eighteenth century, it is because 'man' is a scheming and planning animal that the progress of civilization does not proceed according to a single plan. Now this 'invisible hand' type theory of social progress, one which is often used to justify 'neoliberal' economic and political organization, is something that green political theory questions, but not in the sense of implementing a single plan for social progress. Rather the green case rests on the observation that the various social and ecological problems associated with this view of social progress stem partly from the fact that progress is, increasingly, 'imposed' rather than a spontaneous outcome. Posing the issue of social progress as the outcome of either a 'single plan' (state-imposed) or a market (invisible hand) approach is a false dichotomy, which suggests that the two are mutually exclusive. The correct antonym to an invisible hand view is not a single plan but a deliberative approach to social progress in which the parameters of progress can be decided democratically, i.e. publicly and visibly. The green argument is that the invisible hand approach is no longer appropriate to our current situation. In terms of the relationship between the industrial and French revolutions mentioned above, green political theory holds that social progress both requires, and is constituted by, the management of the former by the latter: that is, the democratic political regulation or governance of socio-economic development. That does not imply the imposition of a single plan, but rather relates to the issue of finding democratic means by which we can choose which forms of social progress we want and which forms we do not. Choosing forms of social progress on this view does not consist in collectively picking one form from amongst a set of possible options. Rather the green argument is that we ought to create democratic and democratically accountable institutions which function to rule out certain forms of progress, those that are unsustainable and/ or parasitic. This is particularly important in the field of technological innovation, where technological developments increasingly raise moral issues concerning the distinction between 'permissible' and 'impermissible' resources and when legitimate human use of the environment becomes unjustified abuse. This negative injunction is also evident in the emphasis on the 'corrective' dimension of the ecological virtue of stewardship as a mode of human interaction which charts a course between the ecological vices of an arrogant anthropocentrism and a submissive ecocentrism, as indicated in Chapter 2. Social progress then is, on the green view, concerned with establishing the (shifting) parameters of a *process* rather than determining the contents of a *product*.[4]

What the green critique of progress represents is a questioning of what one can call the 'Augustinian' view of progress as somehow inevitable or necessary (Nisbet, 1991). The historic evolution of the human species as a linear and ever continuing ascent from poverty, ignorance and fear to affluence, enlightenment and civilization cannot be taken for granted. Perhaps more than anything else green political theory raises fundamental questions concerning the belief in progress understood as the idea that things will inevitably be better in the future. When 'nature' can no longer be taken for granted, in the sense of an independent order and basis for human flourishing, neither can progress.[5] However, not taking progress for granted is not the same as abandoning it as a morally worthy and politically essential social goal. We need to distinguish the concept of progress from particular conceptions.

Rethinking green politics leads to rethinking progress. On the one hand, following Beck (1992), 'progress' needs to mean more than just annual increases in gross national product; it needs to take on a new character and include the democratization of more and more areas of people's lives as being at the heart of a new conception of progress. On the other, progress needs to be separated from the idea that it necessarily requires the domination and exploitation of the non-human world, and an acceptance that our productive relations with the latter are to be judged by normative as well as technical/instrumental criteria. The classical modern view of progress born in the last century needs to be radically rethought. A good example of this ecologically unenlightened conception of progress is T.H. Huxley's view that

> so far from man finding salvation by uncovering the cosmic processes and collaborating with them, it is clear to me that these cosmic forces are hostile and immoral, and it is only by combating them, by man imposing his own moral order on the tiger-rights of a brutal nature that we shall progress. (in Brome, 1963: 11–12)

This old, anthropocentric, arrogant Enlightenment view of progress needs to be critically interrogated, and a new conception of progress in which the aim is to realize human and non-human interests as far as possible is needed. However, having said all that, it would be foolish, not to say politically suicidal, for the green articulation of progress to neglect that even 'orthodox' progress has meant better health, greater life expectancy, and an improvement in people's quality of life. Thus, green politics ought to see itself not as 'anti-progress' but as a politics centrally concerned with extending and redefining progress, which builds upon, while also being critical of, older and now problematic understandings and practices of progress. One model of progress does not fit all. Indeed, the global reach of the Western model of progress (contained in 'development' and 'modernization' theories and practices), sometimes called 'globalization', is the single most responsible cause of much present, past and, if not prevented, future global environmental degradation.

Green Politics and Virtue

This book has defended a conception of green politics the moral basis of which is characterized by a focus on ecological virtue and stewardship. This basis is explicitly anthropocentric, seeing the fulfilment of human interests (particularly human productive or transformative interests) as central and legitimate, if green policy arguments are to have any chance of gaining support from democratic populations. One of the main claims I have made concerns the connection between a particular conception of citizenship as a constitutive aspect of a political process which I have termed collective ecological management (Chapter 5). Central to this is the cultivation of ecological stewardship which consists in the cultivation of ecological virtues and the avoidance of ecological vices. Stewardship as a moral ideal, a form of human excellence, is most clearly expressed within the agricultural context within which this ethical tradition developed. Agricultural stewardship represents a set of interconnected character traits that 'good farmers' would hope to cultivate. Stewardship as 'wise use' is not against human interests, but rather constitutes a mode of action in which future, long-term interests can be safeguarded against the 'temptation' of immediate, short-term ones. Within the family-farm social milieu, in which this version of stewardship originates, relations between humans and nature are characterized by sustainable and symbiotic modes of interaction. While the land and animals are used and consumed, the former is not 'mined' or 'exploited' for short-term profit, nor are the latter treated purely as 'food resources'. In contrast to modern factory farming, 'personal' as well as 'productive' relations exist between animals and humans within the context of agricultural stewardship. As an ideal one can see why it has appealed to many radical green critics of modern, urban life, who see in this ideal a way of directly 'reconnecting' the people and the 'land', which itself is seen as a necessary condition for resolving the 'ecological crisis'. Arguments for returning 'back to the land' have characterized green politics since its origins in the romantic backlash against the industrial society (Gould, 1988).

While one may seek inspiration from this agricultural stewardship tradition, the realities of contemporary Western societies are such that this type of stewardship is not a viable option. Indeed it would be an ecological disaster if urban populations were to return to the land *en masse*. However, as part of a wider process of ecological change in economic and agricultural organization, a selective and voluntary repopulation of rural areas would be desirable for two main reasons, one social and the other ecological. Firstly, such a policy would help to maintain farming as a form of stewardship and farming as a valuable way of life. Secondly, it would be a necessary component of encouraging less industrialized, oil-based forms of food and fibre production. That said, it is naive to suggest, as many greens do, that our ecological problems would be solved if only we were to move out of the cities. Starting from a position in which the majority of Western populations are concentrated in cities and urban areas (those to whom green policy and institutional recommendations must be acceptable), the task facing green politics is how to translate or adapt the moral virtues of a stewardship ethic to a mode of life that, on the face of it, could not be more removed from an agricultural and rural setting. There is also the problem Monbiot identifies of whether 'only those who own or manage the land should be able to determine what happens there' (1997: 4). As argued earlier, the issue is not direct, unmediated experience and management of the environment through ownership and use, but environmental management based on mediated, institutionalized social–environmental interaction. The problem is this: if we reject eco-centrism, but the only acceptable form of an ethics of use is based on modes of human interaction and ways of life which are not the lived experiences of the majority of the population, how can green politics be advanced in a manner which is not (a) a return to a 'pre-modern', agricultural stage of social development, or (b) undemocratic?

Since the agricultural setting within which stewardship developed is no longer available to most people in the Western world, it is clear that virtues and character traits based on it will be difficult to cultivate, to use an appropriate term, within the urban context of modern life. However, just because agricultural stewardship can only be experienced by a minority within an industrial or a post-industrial society, does not mean that it is unimportant in terms of green aims of creating an ecologically rational metabolism between society and environment. Two issues can be mentioned. On the one hand, although the numbers of those working on the land have steadily decreased, along with farming as a proportion of land use, farming still accounts for a large proportion of land use. Therefore, any movement to an ecologically rational form of ecological management would have to take this fact into account. Farmers are, and will continue to be, *de facto* 'ecological stakeholders', if not *de jure* 'agricultural stewards', and thus an important constituency and interest group, as environmental policies from Agenda 21 to EU environmental programmes have acknowledged. On the other hand, while we can think of a 'post-industrial' society, understood as a stage of societal development coming after an industrial phase, the idea of a 'post-agricultural' society is, on the face of it, impossible. While it is of course possible to imagine future social stages where we can synthesize protein from rocks or genetically create food and fibre products in laboratories, for the foreseeable future agriculture, as a mode of human-productive relations with the natural world which requires direct transformative contact with nature, is here to stay. A focus on agriculture is thus not simply for inspirational reasons or for finding resources for a modern stewardship ethic, but also for reasons of practicality in terms of achieving environmental policy aims.

But even given the disproportionate importance of farmers and agriculturists in terms of social–environmental relations, particularly as environmental managers of the land, the fact remains that the 'many' who have to accept and consent to environmental policies, if those policies are to be democratic, live in cities and urban settings. The roles of individuals within urban modes of living are clearly different (in kind and degree) from those modes within the idealized agricultural setting of stewardship. In rethinking green politics, Ferris is surely right to suggest that 'it is time for greens to re-examine their attitude to the urban environment critically and to start proposing policies for promoting sustainable urban development' (1992: 149). Indeed, how to render urban life ecologically sustainable in terms of key areas such as transport, energy consumption, land use, housing, waste generation and disposal, food production and distribution constitute the main challenge in translating green political theory into practical policy outcomes.

Given the urban nature of contemporary life, and discounting green arguments which turn on the supposed 'unnaturalness' of urban living (Goodin, 1992: 52), an *ecological* rather than an *agricultural* form is the most appropriate form that stewardship can take. While the majority of people in modern society have not direct, transformative experiences of nature, this does not mean that the dispositions and attitudes constitutive of stewardship as a mode of action are impossible to cultivate in an urban setting. It was for similar reasons that the emphasis within deep ecology on 'wilderness experience' was questioned as a necessary or desirable component of green moral theory in Chapter 2. Ecological management within agricultural stewardship takes place within the context of farming as a social practice. Urban-based forms of management must necessarily be mediated by social institutions and forms of knowledge not necessarily based on direct experience or ownership of the environment.[6] The most important of these social institutions are the (formal) market economy and the state, while scientific and other forms of ecological knowledge are also important for ecological stewardship. The point is, as green politics has sought to bring home to Western governments and publics, that modern life for the majority of people is not based on direct relations to or experience of the land or the non-human environment. However, the reality is that modern life is not independent of agriculture or, more to the point, the non-human environment. Modern life has not, despite appearances, 'escaped' or 'overcome' a dependence upon the non-human world. As such these various mediating institutions will, and are, central in effecting urban forms of environmental management. Ecological modernization is a good example of an urban-based form of environmental management, as discussed in Chapter 5. The existence of this institutional dimension, that stewardship is ecological, i.e. mediated, not agricultural, in turn means that the 're-embedding' of the economy in society cannot be complete (Chapter 6). Transparency in economic life, while it can be enhanced by turning to non-formal economic spheres such as LETS or commons regimes, is simply impossible to achieve within modern societies. Does this then mean that the 'reintegration' of the human and the natural economy is also impossible? Fortunately, transparency in economic or social life is not a precondition for an ecologically rational metabolism between economy and ecology. While it is true that small-scale, directly democratic bioregional communities do have a lot to offer by way of ecological rationality, they are not the only form that ecologically rational social–environmental arrangements can take. We can establish a degree of harmony with nature and avoid ecological risks, without 'dropping out' of, rejecting or otherwise 'abandoning' contemporary society. Institutional (re)design and (re)orientation can deliver ecologically rational forms of social–environmental interaction, especially when institutional innovation is viewed as a necessary rather than as a sufficient condition.

Now while we may accept the place of institutions in any feasible form of modern, urbanized environmental management, this counts neither against the desirability or necessity for institutions to be supplemented by social practices, nor for a division of management powers in favour of institutional office holders (including existing land owners and owners of economic enterprises) rather than citizens. As far as modern ecological forms of environmental management are concerned, the real issues, as this book has sought to demonstrate, concern the choice of institution (state, market, community), its level (local, national, global), procedures (precautionary principle), management issue (pollution or resource problem), and most importantly, its democratic accountability. And while an institutional framework is a necessary condition for modern environmental management, from a green perspective this needs to be supplemented with a focus on the individual's role in this process. A key aspect of the individual's role is a view of green citizenship as an integrating mode of human interaction. While green citizenship suggests a new combination of rights and duties, its integrative role relates to its function to integrate other modes of human interaction, particularly those of consumption and production, which together constitute 'ecological stewardship' as an overarching mode of interaction with the environment.

It is here that the idea of 'green citizenship' is vital to the green democratic position. On the one hand, green citizenship is argued to be a necessary feature of what I term collective ecological management. The state and formal market cannot do all that is required for an ecologically rational form of environmental management, so citizenship is viewed as an activity, a particular mode of collective and individual action in which the possible ecological vices of consumption may be mitigated or avoided. Citizenship as a mode of character thus transcends the purely 'political' or formal status and legal standing of citizenship, and comes to denote a way of acting which tends towards ecologically rational forms of action. Citizenship within the context of collective ecological management becomes a way of transforming urban dwellers into 'ecological stewards', giving those who may have no direct experience of nature some responsibility for, and democratic input into, managing the metabolism between society and the environment. Hence the vital importance of environmental education to foster this responsibility. One of the most important aspects of green citizenship in this respect is to educate individuals about the dependence of society (which includes them) on the environment, and also the environment's increasing dependence upon and vulnerability to human society (including their actions). As O'Neill suggests,

> Universal indifference to the care and preservation of natural and man-made environments undermines and withers human life and capacities and capabilities for action . . . lives and cultures will remain vulnerable if they depend on environments which, although not damaged, are also not cherished. (1996: 203)

At the same time, the neediness, vulnerability and dependence of humans on one another needs also to be recognized and acknowledged, rather than ignored. In this educative process, scientific knowledge of the world is essential. The 'ethic of use' is thus a particular way of acting in the world, which while being respectful of the non-human world, does not lapse into a submissive 'quietism', criticized in Chapter 2. It accords nature its proper place in the human order of things. On this reading, the emphasis on science within green politics is part of its pedagogic role in shaping this mode of interaction which *integrates* rather than *rejects* material consumption, an attitude towards consumption at odds with the 'standard' green rejection of it, as discussed in Chapter 6. The 'good' towards which the ecological virtues are orientated is pluralistic not singular (O'Neill, 1993). Thus the types of lives that are compatible with stewardship can (within limits) take many forms.[7] Ecological stewardship seeks to promote modes of human life which avoid the extremes (vices) of a 'submissive ecocentrism' and an 'arrogant anthropocentrism'. Just as liberals seek to support and encourage liberal modes of life and discourage illiberal ones, so likewise do greens support ecological modes and discourage unsustainable ones.

In reference to the green critique of a consumption-based economy, the task of green citizenship is to integrate consumption (of formal market commodities) within a stewardship mode of human action, such that it does not become an ecological vice. Consumption is not rejected outright, as it is in many conceptions of green ideology, but rather seen as only one role and activity amongst others of an ecologically rational character. Consumption is not simply an individual activity but a shared social value and a particularly modern human mode of action and experience (Keat, 1994). In raising questions about the status of this social value and its relationship to other social values, such as environmental protection or a concern for future generations, one is also raising issues about the place of material consumption in human life. This is of course a central issue in the green critique of 'progress'. Consumption, particularly within an urban context, can divorce consumption from the ecological processes which have helped to create it. The connection between human consumption and the natural basis of that consumption becomes less clear, as the gap between production and consumption increases. The point is that what needs to be ascertained is that consumption as a valued mode of human action requires seeing how it functions as a mode of human interaction with the various environments humans inhabit (both social and natural). This requires it to be harmonized and balanced within a more expansive mode of interaction denoted by stewardship, in which human productive and consumptive interests can be realized without compromising long-term ecological sustainability. Green politics argues that we should perhaps consume less rather than simply in a 'greener' manner. It is important to remember that 'consumption' here refers to consumption of commodities in the formal market. As was suggested in Chapter 6, one way of consuming less is to engage in ownwork, or LETS-type informal productive activity. Restructuring work so as to allow its internal goods to be realized is a key policy area for green politics in creating a less consumption-driven economy and society (Barry, 1998a).

Stewardship within the modern world cannot refer, as it once did, to knowledge of the natural world gained from direct, productive experience. As a mode of interaction, modern ecological stewardship is premised on the assumption that sufficient knowledge of the world can emerge from the experience of being involved in the political process of collective ecological management. Given that the impact on the environment extends far beyond the agricultural sector to other economic sectors and beyond the 'rural' to include national and global dimensions of environmental change, then environmental management of necessity has to be a political process. Stewardship cannot be left to the agricultural aspect of our relationship to nature since we now interact with it on so many other aspects and levels. A key aspect of this knowledge, stressed throughout, is scientific knowledge, both as a possible metaphysical basis for agreement, but also as a possible basis for policy agreement. At the same time, in keeping with the self-reflexive nature of stewardship, there is also an awareness of the limits to that knowledge. Knowledge is power, but green political theory argues that with power (potentially over all life on earth) comes responsibility. Power without ecological responsibility can lead to an arrogant, self-centred and narrow humanism, while responsibility without power can lead to a submissive timidity in the face of the immensity of the natural order.

Future Issues

In the twenty-first century, a whole series of theoretical and practical issues will face green politics and will be central to its future development. Of course not all of these future issues are new; many are already part of the broad concerns of green politics, and many have been indicated in this book. Some of these include: the development of the South in a just and ecologically sustainable manner in general, and the various modes of resistance and struggle against the universalization of the Western model; the moral, ecological, political and economic concerns around genetic engineering and biotechnology; in the 'developed world', the social and ecological costs of the car; ownership and control relations over the land and over ecological resources and processes more widely; the impacts of 'globalization' and the increasing hegemony of multinational corporations, and the effects of these changes on democratic polities and their implications for the nation-state; the political economy of food production and consumption (especially monocropping and factory farming methods); global climate change; and the numerous other global, regional and local ecological problems. All of these and more will (and are) pressing concerns for green politics. However, there are two future areas which I would like to focus on: the relationship between green politics and justice, and eco-feminist political economy.

As a nascent perspective within political theory, green political theory can be forgiven for not having answers to all or the majority of the standard issues and debates within political theory. Although on a very steep learning curve, the green approach to political theory has many issues, central to the latter, with which to grapple. Of these, the issue of distributive justice is clearly the most important, not least because of the foundational importance of 'scarcity' to both green politics and distributive or social justice. While distributive justice was not directly addressed, some of the parameters, if not the principles, of a putative green theory of justice can be gleaned from the analysis. These parameters relate to the extremely broad scope of a green theory of justice in terms of the recipients of justice. Three new classes of recipients are central to a green theory of justice, and were indicated in the last chapter: non-humans, future generations and foreigners. We can sketch the outlines of a putative green theory of justice in a negative manner. A green theory of justice is one which is characterized by *not* being limited to the distribution of socially produced benefits and burdens within the presently existing human population. More positively, it is concerned with justice between species, between generations and within the present human generation considered globally (Cooper and Palmer, 1995). It is clear that a green theory of justice would have to address the extremely contentious issue of whether the human treatment of non-humans ought to be considered as a matter of justice. While the 'ethics of use' position developed in Chapter 3 could be interpreted as arguing for the extension of justice to our treatment of non-humans, I suggested that this would be difficult to sustain, and indeed unnecessary. While we ought to see our relations with the non-human world in part as moral issues, this is not coextensive with including these relations as strict matters of justice.

The intergenerational dimension of a green account of distributive justice is explicit in the idea of ecological stewardship, and its naturalistic anthropocentric basis developed in Chapters 3 and 5. The global dimension can be found in the distinction between 'ecosphere' and 'biosphere' views of the economy–ecology metabolism. Further 'seeds' can be found in the green critique of economic growth from the point of view of sanctioning economic inequalities, the degree of cultural plurality associated with collective ecological management, and the dominance of the Pareto optimality criterion within economic thought. Social justice is at the heart of sustainability, and many of the green arguments raised in this book have major implications for the conceptualization and implementation of social justice. However, having uncovered some seeds of a green theory of distributive justice, I am happy to leave it to a later time (and to others) to plant and harvest whatever grows from them. I will be more than satisfied if I have helped prepare the soil for this next stage in the evolution of green political theory. In terms of its theoretical development, the relationships between green politics, the global and local dimensions of ecological sustainability, and the global and local dimensions of distributive justice constitute key areas of future research and scholarship.

A related central area for future green political theory is the integration of materialist eco-feminist insights in the development of a practicable and attractive green political economy. Many areas have been omitted or insufficiently discussed in this book, but its lack of analysis of eco-feminist materialism is arguably one of its central weaknesses. While there was some discussion of the central importance of reproductive work and practices in Chapter 6, my firm view is that the theoretical deepening of green politics will turn on the extent to which eco-feminist political economy is made central (Barry, 1998c). While the relationship between green politics and social justice increases its breadth and scope, bringing the life-affirming activities and work of reproduction centre-stage in green politics deepens its analysis. Just as green politics seeks to ecologically 're-embed' human society, likewise green politics must focus on 're-embodying' humanity. Eco-feminism suggests that a less 'arrogant', dominating and patriarchal interaction with the non-human world starts from an explicit recognition of the ineliminable vulnerability, neediness and dependence of human beings in relation to each other and the non-human world (Mellor, 1997; Salleh, 1997). The development and integration of a materialist eco-feminism constitutes a major and pressing issue both theoretically and practically for green politics.

'Only Connect'

The idea and idiom of stewardship as an ecologically virtuous mode of social–environmental interaction are also meant to convey a sense that we need to 'reconnect' with the environment, but not in the way which deep ecologists who advocate 'wilderness experience' suggest. The 'reconnecting' of society and environment, argued for in this book, is one which begins with the recognition that human society is dependent upon and ecologically embedded within the non-human environment, together with the equally important recognition that we are a part of as well as apart from the natural order. As such it seems to me that reconciling society and nature (particularly in relation to justifying human demands on nature) may ultimately require that human ecological stewardship move towards expressing gratitude for and a mindfulness of the origins (both social and natural) of the comforts of modern life. While I do not suggest that we return to the practices of 'thanking God/s' or some deified conception of 'nature', being mindful and not taking for granted the many commodities, services and other goods that modern Western societies currently enjoy, and the comfortable ways of life they support, is important. In this way ecological stewardship expresses a respect as opposed to a reverence for the natural world, as well as ultimately leading to the adoption of a global perspective on social–environmental interaction. In this way, ecological stewardship demands not only avoiding the ecological vice of being ignorant or unmindful of our dependence upon (and responsibility for) the natural environment, but also how 'our' relationship with nature is at the same time a relationship with other human beings. Caring for nature does not (nor should it) imply a decrease in care for humans, but the point about ecological stewardship as a virtuous mode of human–nature interaction is to offer us a mode of being, thinking and acting which both helps us cope with the contingencies of that interaction and helps to make ethically appropriate judgements in cases where the interests of the environment and human society conflict.

Our situation is not so drastic that we require either a new ecocentric consciousness, or anti-democratic forms of environmental regulation; both are equally alien to our culture. Talk of 'ecological crisis' and 'saving the earth', while clearly motivated by a strong concern with our current and near-future ecological predicament, are over-reactions. They fail to focus our attention on the resources within our culture – political, institutional and moral – with which we can seek to cope with environmental risks. Unless we tap into these resources, as I hope to have shown, limits to green political theory will quickly assert themselves, and green politics will become a voice in the wilderness, unable to propose convincing political arguments as to why it should be preserved.

Notes

1 However, the critical reaction to the industrial revolution sometimes went with a critical reaction to the French one as well. A full examination of this dialectic within green political theory would therefore require a full investigation of its conservative historical antecedents and present intimations. The conservative reaction to the Enlightenment is a skeleton which would have to be brought out into the open and addressed as a central part of the continuing attempt to get the green theoretical house in order, as it were. As suggested in the introduction, this book is to be viewed as a contribution to this process.

2 A more extreme version of this position can be found in Lasch's (1991) work, *The True and Only Heaven: Progress and its Critics.*

3 This is, in part, based on the Kantian claim that we can never know the 'thing in itself' since it is always revealed and known as the 'thing in relation to us'.

4 A full elaboration of this point would need to explore the idea of the uniqueness of the present generation (at least in the industrialized world) in terms of its nascent 'historical self-consciousness'. In many respects this present human generation is like no other in that it possesses the technological capacity to irrevocably alter the future state of the planet, but is also the first to be consciously (albeit incoherently) aware of this power. That is,

with an awareness of this power comes a sense of responsibility, and it is this which green politics taps into. Thus green politics can be seen as an attempt to increase this self-awareness and the sense of collective moral responsibility to the future and non-humans which follow from it.

5 And there is growing survey evidence that a majority of citizens in advanced industrial society no longer believe that things will be better in the future (Jacobs, 1996: 2–3).

6 In terms of the ownership of land or the environment, there is a clear movement in favour of limiting (if not eliminating) their private ownership on ecological (and social) grounds. According to Varner, 'Increasingly, taking an ecological view of land forces us to treat it as a public resource that individuals hold only in stewardship (or trust) capacity. Any and every piece of land is involved in diverse ecological processes, and any and every form of land use affects those processes to some extent . . . the eclipse of land as private property is near at hand . . . in this age of ecological literacy we have discovered that land uses depend so heavily on ecological infrastructure – on processes that, if they are property at all, are inherently public property – that it hardly makes sense to conceive of land as private property' (1994: 158). This issue of ownership of and control over the non-human environment is yet another which is likely to be a major area of debate. For a further discussion see Barry (1998b).

7 Though tempting I eschew an argument that can be advanced in favour of ecological limits on lifestyles which follows the liberal view of not tolerating the intolerant. The ecological version would be not to sustain the unsustainable. The interesting point about this is that ecological sustainability is in part a measure of ecological tolerance. If we add to this the earlier point about the relationship between multiculturalism and collective ecological management, the extent of green tolerance of cultural diversity within a green state becomes an even more interesting theoretical question, while remaining of pressing practical concern.

References

Achterberg, W. (1993) 'Can Liberal Democracy Survive the Environmental Crisis?', in A. Dobson and P. Lucardie (eds), *The Politics of Nature: Explorations in Green Political Theory*. London: Routledge.

Achterberg, W. (1996) 'Sustainability, Community, and Democracy', in B. Doherty and M. de Geus (eds), *Democracy and Green Political Thought: Sustainability, Rights and Citizenship*. London: Routledge.

Aguilar, S. (1993) 'Corporatist and Statist Designs in Environmental Policy: The Contrasting Roles of Germany and Spain in the European Community Scenario', *Environmental Politics*, 2 (2).

Agyeman, J. and Evans, B. (1994) 'Introduction', in J. Agyeman and B. Evans (eds), *Local Environmental Policies and Strategies*. Harlow: Longman.

Allaby, M. and Bunyard, P. (1980) *The Politics of Self-Sufficiency*. Oxford: Oxford University Press.

Allison, L. (1991) *Ecology and Utility: The Philosophical Dilemmas of Planetary Management*. Leicester: Leicester University Press.

Altvater, E. (1993) *The Future of the Market: An Essay on the Regulation of Money and Nature after the Collapse of 'Actually Existing Socialism'*. London: Verso.

Anderson, T. and Leal, D. (1991) *Free Market Environmentalism*. Boulder, CO: Westview.

Arendt, H. (1959) *The Human Condition*. Chicago: University of Chicago Press.

Aristotle (1948) *Politics* (translated by E. Barker). Oxford: Oxford University Press.

Ashby, E. (1974) *Reflections on Doom: The Third David Owen Memorial Lecture*. Cardiff: University of Cardiff Press.

Ashford, N. (1993) 'Understanding Technological Responses of Industrial Firms to Environmental Problems', in K. Fischer and J. Schot (eds), *Environmental Strategies for Industry*. Washington, DC: Island.

Attfield, R. and Wilkins, B. (eds) (1992) *International Justice and the Third World: Studies in the Philosophy of Development*. London: Routledge.

Bahro, R. (1994) *Avoiding Social and Ecological Disaster: The Politics of World Transformation*. Bath: Gateway.

Barber, B. (1984) *Strong Democracy: Participatory Democracy for a New Age*. Berkeley, CA: University of California Press.

Barry, J. (1990) 'Limits to Growth', Unpublished MA dissertation, University College Dublin, Dublin.

Barry, J. (1993) 'Deep Ecology and the Undermining of Green Politics', in J. Holder, P. Lane, S. Eden, R. Reeve, U. Collier and K. Anderson (eds), *Perspectives on the Environment: Interdisciplinary Research in Action*. Aldershot: Avebury.

Barry, J. (1994) 'Beyond the Shallow and the Deep: Green Politics, Philosophy and Praxis', *Environmental Politics*, 3 (3).

Barry, J. (1995a) 'Towards a Theory of the Green State', in S. Elworthy, K. Anderson, I. Coates, P. Stephens and M. Stroh (eds), *Perspectives on the Environment 2: Interdisciplinary Research on Politics, Planning, Society and the Environment*. Aldershot: Avebury.

Barry, J. (1995b) 'Nature in Question: What is the Question?', *Environmental Politics*, 4 (1).

Barry, J. (1995c) 'Deep Ecology, Socialism and Human "Being in the World": A Part of Yet Apart from Nature', *Capitalism, Nature, Socialism*, 6 (3).

Barry, J. (1995d) 'Justice, Non-Humans and Green Political Theory: The Inevitability of Pluralism', paper presented at the Department of Social Policy, Royal Holloway, University of London, February.

Barry, J. (1996) 'Sustainability, Political Judgement and Citizenship: Connecting Green Politics and Democracy', in B. Doherty and M. de Geus (eds), *Democracy and Green Political Thought: Sustainability, Rights and Citizenship*. London: Routledge.

Barry, J. (1998a) 'Social Policy and Social Movements: Ecology and Social Policy', in N. Ellison and C. Pierson (eds), *Developments in British Social Policy*. London: Macmillan.

Barry, J. (1998b) 'Greening Liberal Democracy in Theory and Practice: Some Thoughts on the Right, the Good and the Sustainable', paper presented at the European Consortium on Political Research (ECPR) Joint Sessions, Warwick, March.

Barry, J. (1998c) 'Feminism and Socialism (and Ecology) Back Together Again? The Emergence of Ecofeminist Political Economy', *Environmental Politics*, 7 (3).

Barry, J. (1998d) 'Marxism and Ecology', in A. Gamble, D. Marsh and T. Tant (eds), *Marxism and Social Science*. London: Macmillan.

Barry, J. and Proops, J. (1996) 'Local Employment and Trading Systems: Linking Citizenship and Sustainability?' Economic and Social Research Council, End of Project Report.

Beck, U. (1992) *Risk Society: Towards a New Modernity*. London: Sage.

Beck, U. (1995) *Ecological Enlightenment: Essays on the Politics of the Risk Society*. New Jersey: Humanities Press.

Begg, A. (1991) *From Dream to Transition: Green Political Strategy*. Leamington Spa: Ecoprint.

Benson, J. (1978) 'Duty and the Beast', *Philosophy*, 53.

Benton, T. (1993) *Natural Relations: Ecology, Animals and Social Justice*. London: Verso.

Berg, P. (1981) 'Devolving beyond Global Monoculture', *CoEvolution Quarterly*, 32.

Bobbio, N. (1989) *Democracy and Dictatorship: The Nature and Limits of State Power*. London: Verso.

Boehmer-Christiansen, S. (1994) 'The Precautionary Principle in Germany: Enabling Government', in T. O'Riordan and J. Cameron (eds), *Interpreting the Precautionary Principle*. London: Earthscan.

Bookchin, M. (1971) *Post-Scarcity Anarchism*. London: Wildwood.

Bookchin, M. (1980) *Towards an Ecological Society*. Montreal/Buffalo: Black Rose.

Bookchin, M. (1986) *The Modern Crisis*. Philadelphia: New Society.

Bookchin, M. (1990) *Remaking Society*. Montreal and New York: Black Rose.

Bookchin, M. (1991) *The Ecology of Freedom: The Emergence and Dissolution of Hierarchy* (revised edn). Montreal and New York: Black Rose.

Bookchin, M. (1992a) *Urbanization without Cities: The Rise and Decline of Citizenship*. Montreal: Black Rose.

Bookchin, M. (1992b) 'Libertarian Municipalism', *Society and Nature*, 1 (1).

Boulding, K. (1966) 'The Economics of the Coming Spaceship Earth', in H. Jarrett (ed.), *Environmental Quality in a Growing Economy*. Baltimore: Johns Hopkins University Press.

Brennan, A. (1988) *Thinking about Nature: An Investigation of Nature, Value and Ecology*. London: Routledge.

Brennan, A. (1992) 'Environmental Decision-Making', in R. Berry (ed.), *Environmental Dilemmas: Ethics and Decisions*. London: Chapman and Hall.

Brome, V. (1963) *The Problem of Progress*. London: Cassell.

Bromley, D. (1991) *Environment and Economy: Property Rights and Public Policy*. Oxford: Blackwell.

Brubaker, E. (1995) *Property Rights in the Defence of Nature*. London: Earthscan.

Bunyard, P. and Morgan-Grenville, F. (eds) (1987) *The Green Alternative*. London: Methuen.

Cairns, A. and Williams, C. (1985) *Constitutionalism, Citizenship and Society in Canada*. Toronto: University of Toronto Press.

Callahan, D. (1992) 'The Wise Use Movement'. Unpublished MS from the W. Alton Jones Foundation, Washington.

Callicott, J.B. (1982) 'Hume's Is/Ought Dichotomy and the Relation of Ecology to Leopold's Land Ethic', *Environmental Ethics*, 4 (2).

Callicott, J.B. (1992) 'Can a Theory of Moral Sentiments Support a Genuine Normative Environmental Ethic?', *Inquiry*, 35.

Capra, F. (1983) *The Turning Point*. London: Fontana.

Capra, F. (1995) 'Deep Ecology: A New Paradigm', in G. Sessions (ed.), *Deep Ecology for the 21st Century*. Boston and London: Shambala.

Carley, M. and Christie, I. (1992) *Managing Sustainable Development*. London: Earthscan.

Carter, A. (1993) 'Toward a Green Political Theory', in A. Dobson and P. Lucardie (eds), *The Politics of Nature: Explorations in Green Political Theory*. London: Routledge.

Carter, N. (1996) 'Worker Co-operatives and Green Political Theory', in B. Doherty and M. de Geus (eds), *Democracy and Green Political Thought: Sustainability, Rights and Citizenship*. London: Routledge.

Christoff, P. (1996) 'Ecological Citizens and Ecologically Guided Democracy', in B. Doherty and M. de Geus (eds), *Democracy and Green Political Thought: Sustainability, Rights and Citizenship*. London: Routledge.

Clark, M. (1992) 'Tasks for Future Ecologists', *Environmental Values*, 1 (1).

Clark, S. (1979) 'The Rights of Wild Things', *Inquiry*, 22.

Clark, S. (1982) *The Nature of the Beast: Are Animals Moral?* Oxford: Oxford University Press.

Clark, S. (1994) 'Global Religion', in R. Attfield and A. Belsey (eds), *Philosophy and the Natural Environment*. Cambridge: Cambridge University Press.

Clark, S. (1995) 'Enlarging the Community: Companion Animals', in B. Almond (ed.), *Introducing Applied Ethics*. Oxford: Basil Blackwell.

Commoner, B. (1971) *The Closing Circle: Nature, Man and Technology*. New York: Bantam.

Cooper, C. (1989) 'Taking the Greens Seriously: A Libertarian Response to Environmentalism', *Libertarian Alliance: Political Notes*, 35.

Cooper, D. and Palmer, J. (eds) (1995) *Just Environments: Intergenerational, International and Interspecies Issues*. London: Routledge.

Copp, D., Hampton, J. and Roemer, J. (1995) 'Introduction', in D. Copp, J. Hampton and J. Roemer (eds), *The Idea of Democracy*. Cambridge: Cambridge University Press.

Daly, H. (1973a) 'The Steady-State Economy: Toward a Political Economy of Biophysical Equilibrium and Moral Growth', in H. Daly (ed.), *Toward a Steady-State Economy*. San Francisco: Freeman.

Daly, H. (ed.) (1973b) *Toward a Steady-State Economy*. San Francisco: Freeman.

Daly, H. (1985) 'Economics and Sustainability: In Defense of a Steady-State Economy', in M. Tobias (ed.), *Deep Ecology*. San Diego: Avant.

Daly, H. (1987) 'The Steady-State Economy: Alternative to Growthmania'. Unpublished MS.

Daly, H. and Cobb, J. (1990) *For the Common Good: Redirecting the Economy toward Community, the Environment and a Sustainable Future*. London: Greenprint.

de Geus, M. (1991) 'Political Philosophy, Environment and the State', paper presented at the ECPR Workshop on Green Political Theory, Colchester, April.

de Geus, M. (1996) 'The Ecological Restructuring of the State', in B. Doherty and M. de Geus (eds), *Democracy and Green Political Thought: Sustainability, Rights and Citizenship*. London: Routledge.

de-Shalit, A. (1995) *Why Posterity Matters: Environmental Policies and Future Generations*. London: Routledge.

Devall, B. (1988) *Simple in Means, Rich in Ends: Practising Deep Ecology*. Salt Lake, UT: Gibbs Smith.

Devall, B. and Sessions, G. (eds) (1985) *Deep Ecology: Living as if Nature Mattered*. Layton, UT: Peregrine and Smith.

Diamond, C. (1978) 'Eating Meat and Eating People', *Philosophy*, 53.

Die Grünen (1983) *Programme of the German Green Party*. London: Heretic.

Dobson, A. (1989) 'Deep Ecology', *Cogito*, 3 (1).

Dobson, A. (1990) *Green Political Thought*. London: Unwin Hyman.

Dobson, A. (1995) *Green Political Thought* (2nd edn). London: Routledge.

Dobson, A. (1996a) 'Discursive Democracy and the Claims of Green Politics', in B. Doherty and M. de Geus (eds), *Democracy and Green Political Thought: Sustainability, Rights and Citizenship*. London: Routledge.

Dobson, A. (1996b) 'Representative Democracy and the Environment', in W. Lafferty and J. Meadowcroft (eds), *Democracy and the Environment: Problems and Prospects*. London: Edward Elgar.

Dobson, A. (1998) *Justice and the Environment: Conceptions of Environmental Sustainability and Theories of Distributive Justice*. Oxford: Oxford University Press.

Dobson, A. and Lucardie, P. (eds) (1993) *The Politics of Nature: Explorations in Green Political Theory*. London: Routledge.

Dodson-Gray, E. (1981) *Green Paradise Lost*. Wellesley, MA: Roundtable.

Doherty, B. (1992) 'The Fundi–Realo Controversy: An Analysis of Four European Green Parties', *Environmental Politics*, 1 (1).

Doherty, B. (1996) 'Introduction', in B. Doherty and M. de Geus (eds), *Democracy*

and Green Political Thought: Sustainability, Rights and Citizenship. London: Routledge.
Doherty, B. and de Geus, M. (eds) (1996) *Democracy and Green Political Thought: Sustainability, Rights and Citizenship*. London: Routledge.
Dryzek, J. (1987) *Rational Ecology: Environment and Political Economy*. Oxford: Basil Blackwell.
Dryzek, J. (1990) *Discursive Democracy: Politics, Policy and Political Science*. Cambridge: Cambridge University Press.
Dryzek, J. (1995) 'Ecology and Discursive Democracy: Beyond Liberal Capitalism and the Administrative State', in M. O'Connor (ed.), *Is Capitalism Sustainable? Political Economy and the Politics of Ecology*. New York and London: Guildford.
Dryzek, J. (1996) 'Foundations for Environmental Political Economy: The Search for Homo Ecologicus?', *New Political Economy*, 1 (1).
Dryzek, J. (1997) *The Politics of the Earth: Environmental Discourses*. Oxford: Oxford University Press.
Eckersley, R. (1991) 'Green Economics: Overcoming the Credibility Gap', paper presented at the International Conference on Human Ecology, Gotenborg.
Eckersley, R. (1992a) *Environmentalism and Political Theory: Toward an Ecocentric Approach*. London: University of London Press.
Eckersley, R. (1992b) 'Green versus Ecosocialist Economic Programmes: The Market Rules OK?', *Political Studies*, 40 (4).
Eckersley, R. (1993a) 'Disciplining the Market, Calling in the State: Four Competing Models for Integrating the Economy and the Environment', paper presented at the Green Political Economy Workshop, European Consortium for Political Research Conference, Leiden.
Eckersley, R. (1993b) 'Free Market Environmentalism: Friend or Foe?', *Environmental Politics*, 2 (1).
Eckersley, R. (1996a) 'Greening Liberal Democracy: The Rights Discourse Revisited', in B. Doherty and M. de Geus (eds), *Democracy and Green Political Theory: Sustainability, Rights and Citizenship*. London: Routledge.
Eckersley, R. (1996b) 'Liberal Democracy and the Rights of Nature: The Struggle for Inclusion', *Environmental Politics*, 5 (1).
Ehrenfeld, D. (1978) *The Arrogance of Humanism*. Oxford: Oxford University Press.
Ekins, P. (ed.) (1990) *The Living Economy: A New Economics in the Making*. London: Routledge.
Elkington, J. and Burke, T. (1989) *The Green Capitalists: How Industry can make Money and Protect the Environment*. London: Gollancz.
Elliot, R. (1982) 'Faking Nature', *Inquiry*, 25.
Elliot, R. (1994) 'Ecology and the Ethics of Environmental Restoration', in R. Attfield and A. Belsey (eds), *Philosophy and the Natural Environment*. Cambridge: Cambridge University Press.
Elster, J. (1983) *Sour Grapes: Essays in the Subversion of Rationality*. Cambridge: Cambridge University Press.
Elster, J. (1986) 'The Market and the Forum: Three Varieties of Political Theory', in J. Elster and A. Hylland (eds), *Foundations of Social Choice Theory*. Cambridge: Cambridge University Press.
Faber, M., Manstetten, R. and Proops, J. (1992) 'Humankind and the

Environment: An Anatomy of Surprise and Ignorance', *Environmental Values*, 1 (1).

Fairlie, S., Hildyard, N., Lohmann, L. and Sexton, S. (1995) 'Reclaiming the Commons', in J. Lovenduski and J. Stanyer (eds), *Contemporary Political Studies*. Belfast: Political Studies Association.

Ferguson, A. (1966) *An Essay on the History of Civil Society* (1767) (edited by D. Forbes). Edinburgh: Edinburgh University Press.

Ferris, J. (1992) 'Review Article: Green Living', *Environmental Politics*, 1 (1).

Foot, P. (1978) *Virtues and Vices and Other Essays in Moral Philosophy*. Oxford: Basil Blackwell.

Fox, W. (1984) 'Deep Ecology: A New Philosophy of Our Time?', *The Ecologist*, 14 (5/6).

Fox, W. (1990) *Toward a Transpersonal Ecology: Developing New Foundations for Environmentalism*. Boston: Shambala.

Frankel, B. (1987) *The Post-Industrial Utopians*. Oxford: Polity.

Frasz, G. (1993) 'Environmental Virtue Ethics: A New Direction for Environmental Ethics', *Environmental Ethics*, 15 (3).

Freeden, M. (1995) 'Green Ideology: Concepts and Structures'. Oxford Centre for Environment, Ethics and Society, Research Paper no. 4.

Friends of the Earth Europe (1995) *Towards Sustainable Europe*. London: Friends of the Earth Europe.

Galston, W. (1991) *Liberal Purposes: Goods, Duties and Virtues in the Liberal State*. Cambridge: Cambridge University Press.

Garner, R. (1996) *Environmental Politics*. London: Prentice-Hall/Harvester Wheatsheaf.

Geach, P. (1977) *The Virtues*. Cambridge: Cambridge University Press.

Georgescu-Roegen, N. (1971) *The Entropy Law and the Economic Process*. Princeton, NJ: Harvard University Press.

Georgescu-Roegen, N. (1976) *Energy and Economic Myths: Institutional and Analytical Economic Essays*. New York: Pergamon.

Giarini, O. (1980) *Dialogue on Wealth and Welfare*. Oxford: Pergamon.

Goldsmith, E. (1972a) 'Social Disintegration: Causes', in E. Goldsmith (ed.), *Can Britain Survive?* London: Tom Stacey.

Goldsmith, E. (ed.) (1972b) *A Blueprint for Survival*. London: Tom Stacy.

Goldsmith, E. (1988) *The Great U-Turn: De-Industrializing Society*. Bideford: Green Books.

Goldsmith, E. (1991) *The Way: 87 Principles for an Ecological World*. London: Rider.

Goldsmith, E., Hildyard, N., Bunyard, P., McCully, P. Sexton, S. and Fairlie, S. (eds) (1992) 'Whose Common Future?', *The Ecologist*, 22 (4).

Goodin, R. (1985) *Protecting the Vulnerable*. Chicago: University of Chicago Press.

Goodin, R. (1992) *Green Political Theory*. Oxford: Oxford University Press.

Goodin, R. (1996) 'Enfranchising the Earth and its Alternatives', *Political Studies*, 44 (5).

Goodland, R. (1995) 'The Concept of Environmental Sustainability', *Annual Review of Ecological Systems*, 26 (1).

Goodwin, B. (1991) 'Utopianism', in D. Miller et al. (eds), *The Blackwell Encyclopaedia of Political Thought*. Cambridge: Blackwell.

Gordon, J. (1993) 'Letting the Genie Out: Local Government and UNCED', *Environmental Politics*, 2 (4).

Gordon, J. (1994) *Canadian Round Tables and Other Mechanisms for Sustainable Development in Canada*. Luton: Local Government Management Board.
Gorz, A. (1982) *Farewell to the Working Class: An Essay on Post-Industrial Socialism*. London: Pluto.
Gorz, A. (1983) *Ecology as Politics*. London: Pluto.
Gorz, A. (1989) *Critique of Economic Reason*. London: Verso.
Gould, P. (1988) *Early Green Politics*. Brighton: Harvester Wheatsheaf.
Gowdy, J. (1994) 'Progress and Environmental Sustainability', *Environmental Ethics*, 16 (1).
Greco, T. (1994) *New Money for Healthy Communities*. Tucson: author.
Green, T.H. (1974) 'Lecture on Liberal Legislation and Freedom of Contract' (1881), reprinted in J.B. Diggs (ed.), *The State, Justice and the Common Good*. Glenview, IL: Scott, Foresman.
Grey, W. (1986) 'A Critique of Deep Ecology', *Journal of Applied Philosophy*, 3 (2).
Gruen, L. (1993) 'Animals', in R. Goodin and P. Pettit (eds), *A Companion to Political Philosophy*. Oxford: Blackwell.
Gutmann, A. (1987) *Democratic Education*. Princeton, NJ: Princeton University Press.
Habermas, J. (1974) *Legitimation Crisis*. London: Heinemann.
Hampshire, S. (1983) *Morality and Conflict*. Oxford: Blackwell.
Hampshire, S. (1989) *Innocence and Experience*. London: Penguin.
Hardin, G. (1968) 'The Tragedy of the Commons', *Science*, 168.
Hardin, G. (1977) *The Limits to Altruism*. Indianapolis: Indiana University Press.
Hart, H.L.A. (1955) 'Are There Any Natural Rights?', *Philosophical Review*, 64.
Harvey, D. (1993) 'The Nature of Environment: The Dialectics of Social and Environmental Change', *Socialist Register 1993*.
Harvey, D. (1996) *Justice, Nature and the Geography of Difference*. Oxford: Blackwell.
Hayek, F. (1976) *The Road to Serfdom*. London: Routledge and Kegan Paul.
Hayward, T. (1995) *Ecological Thought: An Introduction*. Cambridge: Polity.
Heilbroner, R. (1980) *An Inquiry into the Human Prospect* (2nd edn). New York: Norton.
Held, D. (1987) *Models of Democracy*. Cambridge: Polity.
Held, D. (1991a) 'Between State and Civil Society: Citizenship', in G. Andrews (ed.), *Citizenship*. London: Lawrence and Wishart.
Held, D. (ed.) (1991b) *Political Theory Today*. Cambridge: Polity.
Henderson, H., Lintott, J. and Sparrow, P. (1990) 'Indicators of No Real Meaning', in P. Ekins (ed.), *The Living Economy*. London: Routledge.
Hill, T. (1983) 'Ideals of Human Excellence and Preserving Natural Environments', *Environmental Ethics*, 5 (2).
Hirsch, F. (1977) *Social Limits to Growth*. London: Routledge and Kegan Paul.
Holland, A. (1984) 'On Behalf of Moderate Speciesism', *Journal of Applied Philosophy*, 1 (2).
Holmes, S. (1993) *The Anatomy of Anti-Liberalism*. Cambridge, MA: Harvard University Press.
Holmes, S. (1995) 'Tocqueville and Democracy', in D. Copp, J. Hampton and J. Roemer (eds), *The Idea of Democracy*. Cambridge: Cambridge University Press.
Horkheimer, M. and Adorno, T. (1973) *The Dialectic of Enlightenment*. London: Allen Lane.
Hutchenson, F. (1992) 'Wealth, Waste, Work and Income', paper presented at

the Interdisciplinary Research Network on Environment and Society Annual Conference, Leeds, September.
Illich, I. (1971) *Deschooling Society*. Harmondsworth: Penguin.
Illich, I. (1973) *Celebration of Awareness: A Call for Institutional Revolution*. Harmondsworth: Penguin.
Illich, I. (1974) *Energy and Equity*. London: Marion Boyars.
Illich, I. (1975) *Tools for Conviviality*. London: Fontana.
Illich, I. (1977) 'Disabling Professions', in I. Illich, I. Zola, J. McKnight, J. Caplan and H. Shaiken (eds), *Disabling Professions*. London: Marion Boyars.
Illich, I. (1978) *The Right to Useful Unemployment and Its Professional Enemies*. London: Marion Boyars.
Illich, I. (1981) *Shadow Work*. London: Marion Boyars.
Irvine, S. and Ponton, A. (1988) *A Green Manifesto*. London: Optima.
IUNC, UNEP and WWF (1991) *Caring for the Earth: A Strategy for Sustainable Living*. Gland, Switzerland: The World Conservation Union.
Jacobs, M. (1991) *The Green Economy: Environment, Sustainable Development and the Politics of the Future*. London: Pluto.
Jacobs, M. (1994) 'The Limits of Neoclassicalism: Towards an Institutional Environmental Economics', in M. Redclift and T. Benton (eds), *Social Theory and the Environment*. London: Routledge.
Jacobs, M. (1995) 'Markets, Governance and Sustainability: Steering Ecological Restructuring', paper presented to the Planning Sustainability Conference, Sheffield, 8–10 September.
Jacobs, M. (1996) *The Politics of the Real World*. London: Earthscan.
Jacobs, M. (1997) 'Environmental Valuation: Deliberative Democracy and Public Decision-Making Institutions', in J. Foster (ed.), *Valuing Nature*. London: Routledge.
Jahn, D. (1993) 'Environmentalism: Challenging the Societal Consensus between Labour and Capital?', *Innovation*, 6 (4).
Jansen, A.I. and Osland, O. (1994) 'On the Implications of the Choice of Instruments in Environmental Policy', paper presented at the Green Politics and Democracy Workshop at the ECPR Joint Sessions Conference, Madrid.
Jones, A. (1990) 'Social Symbiosis: A Gaian Critique of Contemporary Social Theory', *The Ecologist*, 20 (3).
Kant, I. (1957) 'Critique of Pure Reason', in T. Greene (ed.), *Kant: Selections*. New York: Charles Scribner's Sons.
Kavka, G. and Warren, V. (1983) 'Political Representation for Future Generations', in R. Elliot and A. Gare (eds), *Environmental Philosophy*. Milton Keynes: Open University Press.
Keane, J. (1988) *Democracy and Civil Society*. London: Verso.
Keane, J. and Owens, J. (1986) *After Full Employment*. London: Hutchinson.
Keat, R. (1994) 'Citizens, Consumers and the Environment: Reflections on *The Economy of the Earth*', *Environmental Values*, 3 (4).
Kemball-Cook, D., Baker, M. and Mattingly, C. (eds) (1991) *The Green Budget: An Emergency Programme for the UK*. London: Greenprint.
Kemp, P. and Wall, D. (1990) *A Green Manifesto for the 1990s*. Harmondsworth: Penguin.
Kerry-Smith, R. (ed.) (1979) *Scarcity and Growth Reconsidered*. San Francisco: Freeman.

Keynes, J.M. (1931) 'Economic Possibilities for our Grandchildren', in his *Essays in Persuasion*. London: Macmillan.

Kidner, D. (1994) 'Why Psychology is Mute about the Environmental Crisis', *Environmental Ethics*, 16 (4).

Kohák, E. (1984) *The Embers and the Stars: A Philosophical Inquiry into the Moral Sense of Nature*. Chicago and London: University of Chicago Press.

Kossoff, G. (ed.) (1992) 'Paper no. 1', *Social Ecology Network*.

Kumar, K. (1978) *Prophecy and Progress*. Harmondsworth: Penguin.

Kymlicka, W. (1990) *Contemporary Political Philosophy: An Introduction*. Oxford: Clarendon Press.

Kymlicka, W. (1993) 'Community', in R. Goodin and P. Pettit (eds), *A Companion to Political Philosophy*. Oxford: Blackwell.

Kymlicka, W. and Norman, W. (1994) 'Return of the Citizen: A Survey of Recent Work on Citizenship Theory', *Ethics*, 104 (1).

LaChapelle, D. (1993) 'Not Laws of Nature, but Li (Pattern) of Nature', in M. Oelschlaeger (ed.), *The Wilderness Condition: Essays on Environment and Civilization*. Washington and Corelo, CA: Island.

LaChapelle, D. (1995) 'Ritual: The Pattern that Connects', in G. Sessions (ed.), *Deep Ecology for the 21st Century*. Boston and London: Shambala.

Lang, P. (1994) *LETS Work: Rebuilding the Local Economy*. Bristol: Grover.

Lang, T. and Hines, C. (1993) *The New Protectionism*. London: Earthscan.

Lasch, C. (1991) *The True and Only Heaven: Progress and its Critics*. New York: Norton.

Latouche, S. (1993) *In the Wake of the Affluent Society: An Exploration of Post-Development*. London: Zed.

Lauber, V. (1978) 'Ecology, Politics and Liberal Democracy', *Government and Opposition*, 13 (2).

Lawrence, D.H. (1968) '*A propos* of Lady Chatterley's Lover', in W. Roberts and H. Moore (eds), *Phoenix II*. London: Heinemann.

Lee, Kai (1993) *Compass and Gyroscope: Integrating Science and Politics for the Environment*. Washington, DC: Island.

Lee, Keekok (1989) *Social Philosophy and Ecological Scarcity*. London: Routledge.

Lee, Keekook (1993a) 'To De-industrialize: Is It So Irrational?', in A. Dobson and P. Lucardie (eds), *The Politics of Nature: Explorations in Green Political Theory*. London: Routledge.

Lee, Keekok (1993b) 'Ecology and the Last-Person Argument', *Environmental Ethics*, 15 (4).

Leopold, A. (1968) *A Sand County Almanac* (1949). Oxford: Oxford University Press.

Lindblom, C. (1977) *Politics and Markets: The World's Political-Economic Systems*. New York: Basic.

Lukes, S. (1984) 'Marxism and Utopianism', in P. Alexander and R. Gill (eds), *Utopias*. London: Duckworth.

MacIntyre, A. (1984) *After Virtue* (2nd edn). Notre Dame, IL: University of Notre Dame Press.

Macpherson, C.B. (1973) *Democratic Theory*. Oxford: Clarendon.

Martell, L. (1994) *Ecology and Society: An Introduction*. Cambridge: Polity.

Mathews, F. (1991) *The Ecological Self*. London: Routledge.

McLaughlin, A. (1994) *Regarding Nature: Industrialism and Deep Ecology*. New York: State University of New York Press.

McLaughlin, A. (1995) 'The Heart of Deep Ecology', in G. Sessions (ed.), *Deep Ecology for the 21st Century*. Boston and London: Shambala.

Meadows, D., Randers, J. and Brehens, W. (1972) *The Limits to Growth: A Report for the Club of Rome's Project on the Predicament of Mankind*. New York: Universe.

Mellor, M. (1992) *Breaking the Boundaries: Towards a Feminist, Green Socialism*. London: Virago.

Mellor, M. (1995) 'Materialist Communal Politics: Getting from There to Here', in J. Lovenduski and J. Stanyer (eds), *Contemporary Political Studies, 1995*. Belfast: Political Studies Association.

Mellor, M. (1997) *Feminism and Ecology*. Cambridge: Polity.

Melucci, A. (1989) *Nomads of the Present: Social Movements and Individual Needs in Contemporary Society*. London: Radius.

Midgley, M. (1983a) *Animals and Why They Matter*. Harmondsworth: Penguin.

Midgley, M. (1983b) 'Duties Concerning Islands', in R. Elliot and A. Gare (eds), *Environmental Philosophy: A Collection of Essays*. Milton Keynes: Open University Press.

Midgley, M. (1992) 'Towards a More Humane View of the Beasts?', in D. Cooper and J. Palmer (eds), *The Environment in Question: Ethics and Global Issues*. London: Routledge.

Midgley, M. (1995) *Beast and Man: The Roots of Human Nature* (revised edn). London: Routledge.

Milbrath, L. (1984) *Environmentalists: Vanguard for a New Society*. Albany, NY: State University of New York Press.

Mill, J.S. (1857) *Principles of Political Economy*, Vol. 2. London: John Parker.

Miller, C. (1988) *Jefferson and Nature: An Interpretation*. Baltimore and London: Johns Hopkins University Press.

Mills, M. (1996) 'Green Democracy: The Search for an Ethical Solution', in B. Doherty and M. de Geus (eds), *Democracy and Green Political Thought: Sustainability, Rights and Citizenship*. London: Routledge.

Mills, S. (1981) 'Planetary Passions: A Reverent Anarchy', *CoEvolution Quarterly*, 32.

Milton, K. (1996) *Environmentalism and Cultural Theory: Exploring the Role of Anthropology in Environmental Discourse*. London: Routledge.

Monbiot, G. (1997) 'Thrown off the Scent', *Guardian*, 16 July.

Morris, D. (1990) 'Free Trade: The Great Destroyer', *The Ecologist*, 20 (5).

Morrison, D. (1987) 'Landscape Restoration in Response to Previous Disturbances', in M. Turner (ed.), *Landscape Heterogeneity and Disturbance*. New York: Springer.

Mulberg, J. (1992) 'Who Rules the Market: Green versus Ecosocialist Economic Programmes', *Political Studies*, 30 (2).

Mulberg, J. (1993) 'Economics and the Impossibility of Environmental Valuation', paper presented at the Interdisciplinary Research Network on Environment and Society Annual Conference, Sheffield.

Naess, A. (1973) 'The Shallow and the Deep, Long-Range Ecology Movement: A Summary', *Inquiry*, 16.

Naess, A. (1984) 'Intuition, Intrinsic Value and Deep Ecology', *The Ecologist*, 14 (5/6).

Naess, A. (1988) 'Deep Ecology and Ultimate Premises', *The Ecologist*, 18 (4/5).

Naess, A. (1989) *Community, Ecology and Lifestyle*. Cambridge: Cambridge University Press.

Naess, A. (1995) 'Self-Realization: An Ecological Way of Being in the World', in G. Sessions (ed.), *Deep Ecology for the 21st Century*. Boston and London: Shambala.

Nash, R. (1989) *The Rights of Nature: A History of Environmental Ethics*. Madison, WI: University of Wisconsin Press.

Neale, A. (1997) 'Organising Environmental Self-Regulation: Liberal Governmentality and the Pursuit of Ecological Modernisation in Europe', *Environmental Politics*, 6: 4.

Nisbet, R. (1991) 'Progress', in D. Miller, J. Coleman, W. Connolly and A. Ryan (eds), *The Blackwell Encyclopaedia of Political Thought*. Cambridge: Blackwell.

North, R. (1995) *Life on a Modern Planet*. Manchester: Manchester University Press.

Norton, B. (1984) 'Environmental Ethics and Weak Anthropocentrism', *Environmental Ethics*, 6 (2).

Norton, B. (1987) *Why Preserve Natural Variety?* Princeton, NJ: Princeton University Press.

Norton, B. (1991) *Toward Unity Among Environmentalists*. Oxford: Oxford University Press.

Norton, B. (1994) 'Economists' Preferences and the Preferences of Economists', *Environmental Values*, 3 (4).

O'Connor, M. (ed.) (1995) *Is Capitalism Sustainable?: Political Economy and the Politics of Ecology*. New York and London: Guildford.

Odum, E. (1983) *Basic Ecology*. Philadelphia: Saunders.

Oelschlaeger, M. (1991) *The Idea of Wilderness*. New Haven, CT: Yale University Press.

Oelschlaeger, M. (ed.) (1993) *The Wilderness Condition: Essays on Environment and Civilization*. Washington and Covelo, CA: Island.

Offe, C. (1984) *Contradictions of the Welfare State*. London: Hutchinson.

Offe, C. and Preuss, U. (1991) 'Democratic Institutions and Moral Resources', in D. Held (ed.), *Political Theory Today*. Cambridge: Polity.

Oldfield, A. (1990) *Citizenship and Community: Civic Republicanism and the Modern World*. London: Routledge.

Olson, M. and Landsberg, H. (eds) (1975) *The No-Growth Society*. London: Woborn.

O'Neill, J. (1993) *Ecology, Policy and Politics: Human Well-Being and the Natural World*. London: Routledge.

O'Neill, J. (1995a) 'Polity, Economy, Neutrality', *Political Studies*, 43 (3).

O'Neill, J. (1995b) 'Public Choice, Institutional Economics, Environmental Goods', *Environmental Politics*, 4 (2).

O'Neill, J. (1995c) 'In Partial Praise of a Positivist: The Work of Otto Neurath', *Radical Philosophy*, 74.

O'Neill, O. (1991) 'Transnational Justice', in D. Held (ed.), *Political Theory Today*. Cambridge: Polity.

O'Neill, O. (1996) *Towards Justice and Virtue: A Constructive Account of Practical Reasoning*. Cambridge: Cambridge University Press.

Ophuls, W. (1977) *Ecology and the Politics of Scarcity*. San Francisco: Freeman.

O'Riordan, T. (1981) *Environmentalism* (2nd edn). London: Pluto.

O'Riordan, T. and Cameron, J. (eds) (1994) *Interpreting the Precautionary Principle.* London: Earthscan.

O'Riordan, T. and Jordan, A. (1995) 'The Precautionary Principle in Contemporary Environmental Politics', *Environmental Values*, 4 (3).

Paehlke, R. (1988) 'Democracy, Bureaucracy, Environmentalism', *Environmental Ethics*, 10 (4).

Paehlke, R. (1989) *Environmentalism and the Future of Progressive Politics.* New Haven, CT and London: Yale University Press.

Pasek, J. (1994) 'International Justice and Environmental Policy', in J. Stanyer and P. Dunleavy (eds), *Contemporary Political Studies.* Belfast: Political Studies Association.

Passmore, J. (1980) *Man's Responsibility for Nature* (2nd edn). London: Duckworth.

Pateman, C. (1970) *Participation and Democratic Theory.* Cambridge: Cambridge University Press.

Pateman, C. (1985) *The Problem of Political Obligation: A Critique of Liberal Theory.* Cambridge: Polity.

Paterson, M. (1995) 'Radicalising Regimes?: Ecology and the Critique of International Relations Theory', in J. MacMillan and A. Linklater (eds), *Boundaries in Question: New Directions in International Relations.* London: Pinter.

Pearce, D. (1992) 'Green Economics', *Environmental Values*, 1 (1).

Pearce, D., Markandya, A. and Barbier, E. (1989) *Blueprint for a Green Economy.* London: Earthscan.

Pearce, D. et al. (1993) *Blueprint 3: Measuring Sustainable Development.* London: Earthscan.

Pepper, D. (1984) *The Roots of Modern Environmentalism.* London, Beckenham: Croom Helm.

Pepper, D. (1993) *Eco-Socialism: From Deep Ecology to Social Justice.* London: Routledge.

Pickering-Francis, L. and Norman, R. (1978) 'Some Animals Are More Equal Than Others', *Philosophy*, 53.

Plumwood, V. (1993) *Feminism and the Mastery of Nature.* London: Routledge.

Polanyi, K. (1957) *The Great Transformation: The Political and Economic Origins of Our Time.* Boston: Beacon.

Popper, K. (1974) *The Open Society and its Enemies*, Vol. I. London: Routledge and Kegan Paul.

Porritt, J. (1984) *Seeing Green: The Politics of Ecology Explained.* Oxford: Basil Blackwell.

Potier, M. (1990) 'Towards a Better Integration of Environmental, Economic and Other Governmental Policies', in N. Åkerman (ed.), *Maintaining a Satisfactory Environment: An Agenda for International Environmental Policy.* Boulder, CO: Westview.

Purdue, D. (1995) 'Hegemonic Trips: World Trade, Intellectual Property Rights and Biodiversity', *Environmental Politics*, 4 (1).

Rawls, J. (1972) *A Theory of Justice.* Oxford: Oxford University Press.

Rawls, J. (1985) 'Justice as Fairness: Political not Metaphysical', *Philosophy and Public Affairs*, 14 (2).

Redclift, M. (1996) *Wasted: Counting the Costs of Global Consumption.* London: Earthscan.

Redclift, M. and Benton, T. (eds) (1994) *Social Theory and the Global Environment*. London: Routledge.
Regan, T. (1982) *All that Dwell Therein: Animal Rights and Environmental Ethics*. Berkeley, CA: University of California Press.
Regan, T. (1983) *The Case for Animal Rights*. Berkeley, CA: University of California Press.
Rennie-Short, J. (1991) *Imagined Country: Society, Culture and Environment*. London: Routledge.
Richardson, D. (1997) 'The Politics of Sustainable Development', in S. Baker, M. Kousis, D. Richardson and S. Young (eds), *The Politics of Sustainable Development: Theory, Policy and Practice within the European Union*. London: Routledge.
Ridley, M. (1995) *Down to Earth*. London: Institute of Economic Affairs.
Roberts, A. (1979) *The Self-Managing Environment*. London: Allison and Busby.
Robertson, J. (1983) *The Sane Alternative: A Choice of Futures* (2nd edn). Author.
Robertson, J. (1985) *Future Work: Jobs, Self-Employment and Leisure after the Industrial Age*. Aldershot: Gower.
Robinson, M. (1992) *The Greening of British Party Politics*. Manchester: Manchester University Press.
Roddick, J. and Dodds, F. (1993) 'Agenda 21's Political Strategy', *Environmental Politics*, 2 (4).
Rodman, J. (1995) 'Four Forms of Ecological Consciousness Reconsidered', in G. Sessions (ed.), *Deep Ecology for the 21st Century*. Boston and London: Shambala.
Rolston, H. (1979) 'Can and Ought We to Follow Nature?', *Environmental Ethics*, 1 (1).
Rolston, H. (1982) 'Are Values in Nature Subjective or Objective?', *Environmental Ethics*, 4 (2).
Rolston, H. (1988) *Environmental Ethics: Duties to and Values in the Natural World*. Philadelphia: Temple University Press.
Rolston, H. (1992) 'Challenges in Environmental Ethics', in D. Cooper and J. Palmer (eds), *The Environment in Question: Ethics and Global Issues*. London: Routledge.
Rothenberg, D. (1992) 'Individual or Community? Two Approaches to Ecophilosophy in Practice', *Environmental Values*, 1 (2).
Routley, R. and Routley, V. (1979) 'Against the Inevitability of Human Chauvinism', in K. Goodpaster and K. Sayre (eds), *Ethics and the Problems of the 21st Century*. Notre Dame, IL and London: University of Notre Dame Press.
Ryle, M. (1988) *Ecology and Socialism*. London: Century Hutchinson.
Sachs, W. (1990) 'Delinking from the World Market', in P. Ekins (ed.), *The Living Economy*. London: Routledge.
Sachs, W. (1995) 'Global Ecology and the Shadow of Development', in G. Sessions (ed.), *Deep Ecology for the 21st Century*. Boston and London: Shambala.
Sagoff, M. (1988) *The Economy of the Earth: Philosophy, Law and the Environment*. Cambridge: Cambridge University Press.
Sale, K. (1980) *Human Scale*. London: Secker and Warburg.
Sale, K. (1984a) 'Mother of All: An Introduction to Bioregionalism', in S. Kumar (ed.), *The Schumacher Lectures*, Vol. 2. London: Blond and Briggs.

Sale, K. (1984b) 'Bioregionalism: A New Way to Treat the Land', *The Ecologist*, 14 (4).
Salleh, A. (1995) 'Nature, Woman, Labor, Capital: Living the Deepest Contradiction', in M. O'Connor (ed.), *Is Capitalism Sustainable? Political Economy and the Politics of Ecology*. New York and London: Guildford.
Salleh, A. (1997) *Ecofeminism as Politics: Nature, Marx and the Postmodern*. London: Zed.
Saward, M. (1993) 'Green Democracy?', in A. Dobson and P. Lucardie (eds), *The Politics of Nature: Explorations in Green Political Theory*. London: Routledge.
Saward, M. (1996) 'Must Democrats be Environmentalists?', in B. Doherty and M. de Geus (eds), *Democracy and Green Political Thought: Sustainability, Rights and Citizenship*. London: Routledge.
Schonfield, A. (1965) *Modern Capitalism*. Oxford: Oxford University Press.
Schumacher, E. F. (1973) *Small is Beautiful: Economics As If People Really Mattered*. London: Abacus.
Seed, J. (1988) 'Introduction', in J. Seed, J. Macey, P. Fleming and A. Naess, *Thinking Like a Mountain: Towards a Council of All Beings*. London and New York: New Society/Heretic.
Seed, J., Macey, J., Fleming, P. and Naess, A. (1988) *Thinking Like a Mountain: Towards a Council of All Beings*. London and New York: New Society/Heretic.
Selman, P. (1994) 'Canada's Environmental Citizens: Innovation and Partnership for Sustainable Development', *British Journal of Canadian Studies*, 9 (1).
Sen, A. (1981) *Poverty and Famine: An Essay on Entitlement and Deprivation*. Cambridge: Clarendon.
Sessions, G. (1993) 'Ecocentrism, Wilderness, and Global Ecosystem Protection', in M. Oelschlaeger (ed.), *The Wilderness Condition: Essays on Environment and Civilization*. Washington and Covelo, CA: Island.
Sessions, G. (1995a) 'Introduction to Part Six', in G. Sessions (ed.), *Deep Ecology for the 21st Century*. Boston and London: Shambala.
Sessions, G. (1995b) 'Deep Ecology and the New Age Movement', in G. Sessions (ed.), *Deep Ecology for the 21st Century*. Boston and London: Shambala.
Shepard, P. (1993) 'A Post-Historic Primitivism', in M. Oelschlaeger (ed.), *The Wilderness Condition: Essays on Environment and Civilization*. Washington and Covelo, CA: Island.
Shiva, V. (1988) *Staying Alive: Women, Ecology and Development*. London: Zed.
Shragge, E. (ed.) (1993) *Community Economic Development: In Search of Empowerment*. Montreal: Black Rose.
Singer, P. (1979) 'Not for Humans Only: The Place of Non-Humans in Environmental Issues', in K. Goodpaster and K. Sayre (eds), *Ethics and the Problems of the 21st Century*. Notre Dame, IL and London: University of Notre Dame Press.
Singer, P. (1990) *Animal Liberation* (2nd edn). London: Jonathan Cape.
Skolimowski, H. (1993) *A Sacred Place to Dwell: Living with Reverence upon the Earth*. Dorset: Element.
Slote, M. (1993) 'Virtue', in P. Singer (ed.), *A Companion to Ethics*. Cambridge: Blackwell.
Splash, C. and Hanley, N. (1995) 'Preferences, Information and Biodiversity Preservation', *Ecological Economics*, 12 (3).
Spretnak, C. and Capra, F. (1985) *Green Politics: The Global Promise*. London: Paladin.

Stone, C. (1974) *Should Trees Have Standing?* Los Altos, CA: Kaufman.
Stretton, H. (1976) *Capitalism, Socialism and the Environment.* Cambridge: Cambridge University Press.
Sunstein, C. (1995) 'Democracy and Shifting Preferences', in D. Copp, J. Hampton and J. Roemer (eds), *The Idea of Democracy.* Cambridge: Cambridge University Press.
Tännsjö, T. (1992) *Populist Democracy: A Defence.* London: Routledge.
Taylor, C. (1989) *Sources of the Self: The Making of the Modern Identity.* Cambridge: Cambridge University Press.
Taylor, M. (1976) *Anarchy and Co-operation.* London: Wiley.
Taylor, M. (1982) *Community, Anarchy and Liberty.* Cambridge: Cambridge University Press.
Taylor, M. (1987) *The Possibility of Co-operation.* Cambridge: Cambridge University Press.
Taylor, P. (1986) *Respect for Nature: A Theory of Environmental Ethics.* Princeton, NJ: Princeton University Press.
Thompson, P. (1995) *The Spirit of the Soil: Agriculture and Environmental Ethics.* London: Routledge.
Tilly, C. (1992) *Coercion and Capital: European States 900–1992.* Cambridge: Blackwell.
Tocqueville, A. de (1956) *Democracy in America* (edited by R. Heffner). New York: Mentor.
Toffler, A. (1970) *Future Shock.* London: Pan.
Tönnies, F. (1957) *Community and Society.* New York: Harper and Row.
Trainer, F. (1985) *Abandon Affluence!* London: Zed.
UNCED (1992) *Earth Summit 1992.* London: Regency.
Vadnjal, D. and O'Connor, M. (1994) 'What is the Value of Rangitoto Island?', *Environmental Values,* 3 (4).
Van der Straaten, J. (1992) 'The Dutch National Environmental Policy Plan: To Choose or To Lose', *Environmental Politics,* 1 (1).
Van Parijis, P (ed.) (1992) *Arguing for Basic Income.* London: Verso.
Varner, G. (1994) 'Environmental Law and the Eclipse of Land as Private Property', in F. Ferre and P. Hartel (eds), *Ethics and Environmental Policy: Theory Meets Practice.* Athens, GA: University of Georgia Press.
Vincent, A. (1993) 'The Character of Ecology', *Environmental Politics,* 2 (2).
Wall, D. (1990) *Getting There: Steps to a Green Society.* London: Greenprint.
Wall, D. (1994) 'Towards a Green Political Theory: In Defence of the Commons?', in J. Stanyer and P. Dunleavy (eds), *Contemporary Political Studies, 1994.* Belfast: Political Studies Association.
Walker, K. (1989) 'The State in Environmental Management: The Ecological Dimension', *Political Studies,* 37 (1).
Watson, R. (1983) 'A Critique of Anti-Anthropocentric Biocentrism', *Environmental Ethics,* 5 (3).
Weale, A. (1992) *The New Politics of Pollution.* Manchester: Manchester University Press.
Weber, M. (1968) *Economy and Society.* New York: Bedminster.
Weston, J. (ed.) (1986) *Red and Green: The New Politics of the Environment.* London: Pluto.
Westra, L. (1989) 'Ecology and Animals: Is There a Joint Ethic of Respect?', *Environmental Ethics,* 11 (3).

White, L. (1967) 'The Historic Roots of Our Ecologic Crisis', *Science*, 155.
White, S. (1995) 'Liberal Equality, Exploitation and the Case for an Unconditional Basic Income', in J. Lovenduski and J. Stanyer (eds), *Contemporary Political Studies, 1995*. Belfast: Political Studies Association.
Whitebook, J. (1981/2) 'Saving the Subject: Modernity and the Problem of the Autonomous Individual', *Telos*, 50.
Whiteside, K. (1994) 'Hannah Arendt and Ecological Politics', *Environmental Ethics*, 16 (4).
Williams, B. (1992) 'Must a Concern for the Environment be Centred on Human Beings?', in C. Taylor (ed.), *Ethics and the Environment*. Oxford: Corpus Christi College.
Williams, C. (1995) 'Trading Favours in Calderdale', *Town and Country Planning*, 64.
Windass, S. (1976) 'An Alternative Society', *Yale Review*, 65 (4).
Wissenburg, M. (1993) 'The Idea of Nature and the Nature of Distributive Justice', in A. Dobson and P. Lucardie (eds), *The Politics of Nature: Explorations in Green Political Theory*. London: Routledge.
Wissenburg, M. (1998) *The Free and the Green Society: Green Liberalism*. London: University College London Press.
World Commission on Environment and Development (1987) *Our Common Future*. London: Oxford University Press.
Worster, D. (1994) *Nature's Economy: A History of Ecological Ideas* (2nd edn). Cambridge: Cambridge University Press.
Yearley, S. (1991) *The Green Case*. London: Harper Collins.
Young, S. (1992) 'The Different Dimensions of Green Politics', *Environmental Politics*, 1 (1).
Young, S. (1993) *The Politics of the Environment*. Manchester: Baseline.
Young, S. (1994) 'Lasooing the Oil Tanker: The Liberal Democratic State Confronts the Environmental Crisis', paper presented at the ECPR Joint Sessions Workshop, 'Green Politics and Democracy', Madrid, April.
Zimmerman, M. (1993) 'The Blessing of Otherness: Wilderness and the Human Condition', in M. Oelschlaeger (ed.), *The Wilderness Condition: Essays on Environment and Civilization*. Washington and Covelo, CA: Island.

INDEX